International Series in Advanced Management Studies

More information about this series at http://www.springer.com/series/15195

Elena Candelo

Marketing Innovations in the Automotive Industry

Meeting the Challenges of the Digital Age

Società Italiana di
MANAGEMENT

Elena Candelo
Department of Management
University of Turin
Turin, Italy

ISSN 2366-8814   ISSN 2366-8822   (electronic)
International Series in Advanced Management Studies
ISBN 978-3-030-15998-6   ISBN 978-3-030-15999-3   (eBook)
https://doi.org/10.1007/978-3-030-15999-3

Library of Congress Control Number: 2019934784

This Springer imprint is published by the registered company Springer Nature Switzerland AG
The registered company address is: Gewerbestrasse 11, 6330 Cham, Switzerland

If you want to do something new, you have to stop doing something old

Peter Drucker

Preface

Since the outset, marketing strategies in the car industry have been evolving constantly. The best manufacturers have managed to adjust their marketing strategies, often quickly, in response to variations in technology, demand and competition.

In the marketing literature, the evolution of marketing strategies has been analysed mainly across the phases of production orientation (or production approach), sales orientation (or sales approach) and marketing orientation (or marketing approach, sometime defined as the result of the marketing revolution).

This book will provide an in-depth discussion of these approaches, while placing a broad emphasis on the impact of technological innovation, and changes in buyer behaviour and the structure of competition on the industry.

Why study the evolution of marketing strategies in the car industry, specifically? The car industry is a context with all the essential elements for researching how changes in technology, buying behaviour and competition have influenced marketing strategies. It is a mature industry that makes use of well-established technology, which is vulnerable to the unexpected rise and fall of demand, prone to overcapacity, and has high barriers to both entry and exit and a widespread, increasing focus on short-term financial returns.

Many other industries, such as aerospace, agricultural machinery and industrial equipment, have faced the same kinds of variations and challenges: frequent technological innovation generating new paradigms, cyclical markets with deep fluctuations in demand, changes in buyer behaviour and relentless new models of competition. However, the advantage of studying the car industry is that there is a large amount of up-to-date information about many firms with different dimensions, financial resources and organisational structures.

This large amount of up-to-date information about many different companies (carmakers, dealers, advertising agencies, suppliers, logistics firms and so on) from various countries have been collected and analysed using a mixed qualitative methodology which leveraged on several data sources.

First, a broad literature review about the evolution of marketing strategies and innovation management in the automotive industry was performed. This review was enriched with a personal reworking of the information to identify the links between innovation and changes in marketing and industry stakeholders.

The second step consisted in gathering information from the carmakers' archival data (press releases and corporate documents), annual reports, institutional websites, top managers' public speeches and direct company observations, including visits to several manufacturing plants, units and research centres to reconstruct the impact and importance of technological innovation in relation to marketing strategy development.

Subsequently, quantitative and qualitative data were collected through direct interviews with corporate managers of various stakeholders. A number of criteria were used to select the respondents: extensive tenure in the automotive industry (carmakers or others companies), direct interaction with technological innovation and/or marketing activities, and functional and hierarchical relevance and diversity. Thus, knowledgeable informants were selected as they are, were or had been directly involved (in various periods) in the phenomena being investigated, thus providing direct experiences and perceptions. Semi-structured, open-ended interviews were used. Using this type of research instrument allows for the possibility that the discussion might also include areas not planned by the interviewer. The entire data collection phase, including both interviews and the various aforementioned information sources, lasted three years (2015–2018).

The ultimate objective is to foresee, on the basis of past and present indications, the likely future developments in competition in the car industry (consolidation among firms, market fragmentation, disruptions by newcomers) and in marketing strategies. This book focuses on two factors: progress in technological innovation (particularly the advances of the digital revolution) and current changes in consumer and organisational buying behaviour.

In relation to the car industry, this publication has four main objectives:

1. To retrace/run through the evolution of marketing strategies, and of the related business models of which such strategies were an integral part.
2. To identify the factors that influenced this evolution and carmakers' responses during the various stages.
3. To situate the stages of the evolution of marketing strategies in relation to technological innovation in the industry, changes in consumer and organisational buying behaviour, and alterations to the structure of the sector.
4. To explore the likely future evolution of marketing strategies in the car industry and the main innovative factors that could impact them.

This research looks at three partially overlapping phases/periods in the evolution of innovation and marketing, each of which has its own drivers and marketing strategies: (1) the mechanical age (from its early years to the late 1960s); (2) the electronic and software age (from the 1970s to the mid-1990s); and (3) the digital age (from the late 1990s onward).

Part I and Part II deal with the mechanical age, covering a period stretching from its early days to roughly the end of the 1960s.

In both Europe and the USA, the first cars were sold to rich individuals who were passionate about innovation and set on standing out and showing off their status. Vehicles were built at the specific request of individual buyers, who generally purchased the chassis from one manufacturer and the body from another. Since they were built by hand, also using components from other products, they were almost all different from one another. In the early days of the car industry, manufacturers earned recognition through success at competitions with long-distance races which selected the most robust and fastest cars.

A first turning point occurred in the USA in the first decade of the last century. Henry Ford understood that the potential market for a means of transport affordable by the masses was enormous. The price could be kept down by producing large volumes of identical cars to take advantage of economies of scale and reduce the product costs per unit. Mass production and mass marketing were born. In those years, the main factors that influenced marketing were the emergence of new technologies, economic development (constantly growing average income per capita), and the advent of mass consumption, as well, first and foremost, as the capacity of new entrepreneurs to understand the trends of potential demand and coordinate the use of the technological and management knowledge available in various sectors.

Towards the end of the 1920s, a second turning point occurred in the evolution of car marketing. General Motors (GM) overtook Ford in terms of sales volumes. It offered more attractive cars. Contrary to Ford's approach, GM concentrated much more on the styles, colours and external appearance of its vehicles. GM's marketing strategy was geared at segmentation: "a car for every purse and purpose". The strategy was based on the class divisions that were coming to shape American society. For the different social classes, GM offered different products at different prices, seeking to attract potential clients from different segments through different advertisements. It offered high performance for Pontiac, cutting-edge technology for Oldsmobile, reliability for Buick, and a symbol of wealth and power for Cadillac owners.

In Europe and Japan, the history of the industry and the evolution of marketing strategies took a very different path from that of the USA.

In western Europe, one cannot talk about marketing in the modern sense of the term until the 1920s. Cars were mainly produced on a commission basis, at the request of audacious rich individuals who mostly wanted to show off their high social status. The best manufacturers were selected through competitions involving endurance and speed. Such races were the main promotional tool. European supremacy in automotive constructions was confirmed by the fact that rich Americans purchased vehicles in Europe to compete in races in their own country. In 1905, a Fiat 90 cv won the world record over a mile in Long Island.

Between the two world wars, the evolution of marketing strategies was considerably behind compared to developments in the USA. Various factors contributed to this. First and foremost, the middle classes did not have the necessary resources to purchase new cars and the few customers that could afford them called for powerful cars possibly with unique characteristics. It was customary for them to purchase the chassis from one manufacturer and the body from another, requesting innovation and distinction. Secondly, even in the biggest markets, car sales were somewhat modest. No market in Europe, except for Great Britain, was sufficiently large to justify major investments in assembly facilities. Lastly, Europe was divided into several national markets, separated by barriers of various types. After the Second World War, several years of closed borders and governments' desire to defend national identities and industrial traditions contributed to maintaining marked differences between the products and styles of European car manufacturers for a long time yet.

In the 1950s, convergence towards similar forms of distribution began in the main European markets. The introduction of regulations set by the European Community contributed to this. The distribution model that, with few variations, still exists in western Europe was born. Starting with the need to offer clients a guarantee that repairs carried out in dealers' garages were safe, the national authorities permitted manufacturers to assign exclusive sales areas to their dealers and to require the use of original spare parts and standards for premises used for sales purposes (design, layout and size), staff qualifications, equipment at repair shops and the key elements for services offered to clients. In practice, the same dealer could not serve more than one manufacturer, nor could cars be sold through other distribution channels. Structures nonetheless remained very different from one nation to the next in terms of the concentration of dealers.

In the 1960s, substantial differences began to emerge in the main markets between the marketing strategies of three types of manufacturers: mass market; "high-end"; and niche (sports cars and luxury cars). Mass market manufacturers chose mass marketing strategies based on the expansion of their share of the market to create and sustain economies of scale and scope, maximum standardisation of parts and components, a high degree of modularity, and low and medium price ranges geared at targets with the greatest potential for overall demand. Fiat, Renault and Opel were among the most active.

Based on "high-end" (or "premium") strategies, on the other hand, there was the search for differentiation, a prerequisite for charging a "premium price". The greater the differences perceived by the buyer, the more the manufacturer could establish demand, maintain customer loyalty and charge a premium price. Many of the elements for creating differentiation were associated with marketing, from the quality of pre- and after-sales services to the brand image. The "premium" German brands began to emerge: BMW, Mercedes Benz, Audi and Porsche.

Niche manufacturers had inherited the old artisanal production methods, having survived a strict selection process (including Aston Martin, Ferrari, Lamborghini, Jaguar, Rolls Royce and Bentley). They managed to impose their "product concepts" on the market and to preserve their clients' loyalty they kept it unchanged. They controlled consumer access by increasing prices and limiting production volumes to heighten the sense of exclusivity. There were drawbacks, however, as illustrated by the fact that almost all niche manufacturers either lost their independence or left the market.

In Japan, car distribution was traditionally controlled directly by manufacturers and occurred primarily on a door-to-door basis. This came at a high cost for the manufacturer, but with the advantage of increasing their capacity to boost customer loyalty. The turning point came about in the early 1960s, when the government, through the Ministry of Industry ("MITI"), decided to strengthen economic development by investing heavily in certain sectors deemed strategic, including the car industry. Economic growth brought about a rise in the demand for automobiles. At the end of the decade, Japan joined the group of countries with major car construction companies.

The marketing strategies resulting from the change differed greatly from the ones in use in the USA and western Europe. As well as economic growth and technological progress (lean production, the third paradigm), they were influenced by three factors.

The first factor was direct rivalry between individual products and manufacturers within segments. In practice, these segments were isolated from one another. The "product concepts" of the same firm changed from one segment to another.

The second factor was a considerable surge in competition in the internal market. This happened for two reasons: first and foremost, many companies were attracted by the rapid development of sales (in 1950, no fewer than eleven bought mass production plants); and secondly, the instability of "product concepts" mentioned earlier had an impact, rendering any position acquired by a company on the market constantly vulnerable.

The third factor was related to car buyers, who were mainly first-time buyers with limited experience of purchasing such products. They were very attracted by the launch of new products. Given that the competition between manufacturers offered a continuous stream of new products, buyers quickly abandoned one manufacturer in favour of another, often choosing radically different "product concepts". This instability of customer tastes generated instability in demand trends, shortened the product life cycle and hampered companies' medium- to long-term planning.

Part III deals with the electronic and software age. This was a period of profound change in marketing strategies spurred on by innovative technology and two oil crises.

The 1970s and the 1990s saw extraordinary technological innovation in the car industry, which influenced marketing strategies. Electronic engine control units and fuel injection systems were introduced in the 1960s and 1970s, as well as the first airbags and audio systems capable of fully reproducing the quality of home systems. From the mid-1990s, cars began to incorporate autonomous features linked by software. The software age brought about a stream of enhancements from collision-warning systems to the first models with an adaptive cruise control system. For decades, carmakers had been accustomed to adapting their strategies based on the introduction of new technologies. In these two decades, however, the change factor was once again of an economic nature. Many rules were broken.

The age of electronics and software also accorded suppliers greater power and a more important role. Although suppliers create 70–75% of the value of a car, even the largest are little-known. The greatest visibility in the car industry is held by carmakers, called OEMs, who assemble vehicles and sell them to customers. In the mechanical age, companies that assembled cars from thousands of parts had more power than their many small suppliers. In the electronic and software ages, the balance of power changed. Assemblers were less able to control and squeeze major suppliers. As cars became electronic, and later with software, suppliers started to acquire a greater and more powerful role.

The 1970s witnessed a change in the old mass production and mass marketing panorama, based on economies of scale and scope and a focus on efficiency through market stability and control of its variables. Homogeneous markets and the prevalence of standardised products with long life cycles were a thing of the past; instead, markets became heterogeneous and fragmented, and product life cycles shortened. Companies offered a greater variety of products and other elements of the marketing mix through flexibility and a fast response to customers, who requested both standardised products, attracted by their low cost, and differentiated products for which they were willing to pay a premium price.

Many reasons for this trend have been identified: technological innovation; intense competition; overcapacity; and an increasing, widespread focus on short-term financial returns in the industry. The main event triggered by this sequence of changes and innovations in the car industry was of an economic nature, however: the increase in the price of oil.

The two oil crises of the 1970s are considered as a turning point, instigating a decisive change. They sped up a change that was partly already underway. Not only did the way of competing change, but so did society, markets, technologies and customers. The 1970s was also a period of rapid inflation in the USA and western Europe. As prices rose, in some years at "double digits", the adjustment of car sales prices created serious difficulties for marketing strategies in a phase in which demand was constrained by a reduction in the real income of consumers. In periods of inflation, several elements of the marketing mix changed: the "product concepts" (new engines, aerodynamics for greater fuel efficiency, tyres with less resistance to rolling); product mixes (trading down versus small cars); pricing policies (not based on historic costs but rather on restocking costs); and advertising (which needed to highlight product life and the price/value ratio, and tone down the image of a luxury product). The general trend was to move towards downsizing.

In the USA, the market share of the Big Three decreased. When the government laid down regulations geared at reducing gasoline consumption, Japanese and European manufacturers found themselves in an advantageous position. Traditionally, their cars were smaller and, above all, more fuel efficient. American carmakers, particularly GM, responded to the increase in gasoline prices incorrectly. To address the problem, they tried brushing up old small car projects, which they continued to propose alongside the "old" gas-guzzling vehicles. This marketing strategy created confusion among consumers, who preferred to move to the product lines of Japanese manufacturers.

In Japan, the oil crisis of the early 1970s occurred at a time when the strategy deployed by manufacturers was changing once again. As the internal market was already almost saturated, even with the government's assistance, Honda, Nissan and Toyota increased their investments directed at increasing their penetration of overseas markets. Most of the exports went to the USA. The success derived not only from the increased fuel efficiency of the vehicles, but also from a clever marketing strategy based on three elements: attacking the market "blind side" in the segments overlooked by American manufacturers; "listening to the voice of the customers" so as to interpret changes in consumer behaviour in a timely manner; and patiently building up a distribution network. This was all combined with the capacity to gain recognition for product quality, an attentive after-sales service, active and passive vehicle safety, and a reduction in harmful emissions into the atmosphere. The reaction by American manufacturers to constrain Japanese advances was disastrous. They requested, and obtained, the intervention of the government, which set limits on the number of vehicles imported from Japan. Japanese manufacturers responded by reducing small car exports and concentrating their efforts on vehicles that offered higher profits, including the "premium brands": Lexus (Toyota), Infiniti (Nissan) and Acura (Honda).

In Europe, the two oil crises of the 1970s had two profound, long-lasting effects on marketing strategies. The first was that consumers became more cost-conscious. After a phase in which they were surprised and hesitant, consumers soon realised that the decline in their purchasing power was destined to be long-lasting. Consequently, carmakers were forced review the principles of the entire marketing mix. The second effect concerned a group of manufacturers that laid the foundations for dominating the industry in subsequent decades (in terms of profitability and technological innovation). The sports cars and luxury cars that had dominated their respective niches in the preceding decades were overcome by the crisis. As well as their high consumption levels, it became clear that they were behind in terms of R&D and safety. After a period of "wait and see", the German manufacturers Audi, BMW, Mercedes and Porsche succeeded in positioning their "premium brands" in the top-end segments of the car industry through new technologies, a strong image and appropriate pricing policies.

With the change in consumer behaviour and more intense competition, the capacity to differentiate products, which had achieved relative stability and efficiency in the previous two decades, weakened rapidly and continuously. In particular, segmentation techniques lost some of their efficiency. However, progress in electronics came to the rescue. In the previous decades, marketers had succeeded in narrowing their focus and selecting elements to produce cars designed for specific segments. During the late 1970s, thanks to new information technologies, it was possible to carve out a segment tailored to the individual. The expression "segment of one" was coined to designate the capacity to track, know and understand individual customer behaviour. Thanks to advances in electronics, there were greater opportunities for capturing data, and both storage costs and large database management costs decreased drastically.

Also as a result of these advances, marketing strategies underwent a further development in that the major carmakers started to extend the perimeters of their brand portfolios. Driven by the need to bring down average costs, they entered new segments by further expanding upon their models.

The sale of cars to organisations in addition to individuals dates back to the early days of the industry. Rental fleets entered the car industry in the first decade of the last century. In the 1970s, marketing geared at organisations underwent a new development driven by the growth in overall demand for cars. In the meantime, the types of clients had multiplied—in addition to rental fleets, there were also leasing firms and networks of dealers—and in this case, too, progress in electronics had a significant impact (both on relations between organisations and their clients and on those between organisations). The relationship with rental fleets became integral to manufacturers' development strategies. With shares sometimes even surpassing 50% of the total sales, for a manufacturer, stabilising relations with the fleets meant stabilising programmes relating to production, the development of new products, and, in particular, marketing.

Towards the 1990s, certain trends that had developed in the previous decades acquired greater force and drove further changes in marketing strategies. Limits to the mass production process emerged, the homogeneity of the markets was once again reduced, and demand became more unstable. The saturation of the major markets contributed to a shift in negotiating power between the parties involved in the transactions. There was a transition from a sellers' market to a buyers' one, and from markets "ruled" by sellers to ones "ruled" by buyers. The introduction of new technologies and production methods also played a part in reducing the stability of demand in the markets. Thanks to lean production, which allowed for lower costs at smaller volumes with higher quality standards and computer-integrated manufacturing, it became more economical to design and produce a greater variety of products in the automotive industry.

The background of the automotive industry changed in less than a decade. The old rules of mass production, based on stability and control of demand, were no longer sustainable. In place of homogeneous markets, standardised products and long product life cycles, heterogeneous and fragmented markets, a variety of products and shortening product life cycles emerged. The carmakers responded through "mass customisation". While in mass production low costs are achieved primarily through economies of scale, in mass customisation they are principally obtained through economies of scope. In other words, while in mass production the lowest unit costs for single products or services are obtained through high volumes, in mass customisation they are achieved by distributing fixed costs over a large variety of products and services. While mass production has made it possible to increase demand by offering products at low costs, mass customisation has generated similar results through an increased capacity to offer many customers exactly what they ask for.

In the mid-1990s, in Western markets, the evolution of industries and technologies was accompanied by an emerging shift in competitive advantages from "upstream" activities towards "downstream" ones. Info-intermediaries emerged, such as Autobytel.com, which took power away from carmakers (OEMs) in distribution and granted more power to consumers. The centre of gravity shifted towards the "downstream" area, and therefore also towards marketing, requiring many principles to be revised. This also paved the way for the entry of new competitors onto the market.

The car industry resisted these threats. It was more resilient to the value migration than expected, but for the management of major carmakers it was clear that the change called for a response to important strategic questions such as: "Where will profits emerge in the new digital infrastructure?" and "How can we reshape our business to take advantage of the new opportunities better and faster than competitors?". The responses to these questions all involved the incorporation and assimilation of new digital technologies.

On the threshold of the early years of the new millennium, the advances towards a transformation in the automotive industry were already clear. Information technology had begun to be deeply integrated as a tool in marketing research, and in the other main functions: from sourcing to product design, from logistics to manufacturing and from marketing to after-sales services. Products could be better designed and manufactured. Customers' reactions and expectations could be quickly understood and analysed. Cars became more reliable, maintenance less frequent, and repairs rarer.

Part IV deals with the digital age, a period characterised by radical technological innovation with a profound change in strategies in the industry.

On the threshold of the new millennium, the digital transformation is breaking the rules of marketing strategies. The technological revolution in the automotive industry is so radical that the word disruption is frequently used. In the car industry, potential disruption is generated by the combined effect of the progress of more digital technologies, the use of platforms (Uber, Lyft, Didi), electromobility (Tesla) and autonomous driving (Apple and Google). Companies that fall victim to disruption will not necessarily disappear. The self-driving car pioneered by Google might become the main mode of transport within two to three decades. This could revolutionise transport, but the traditional car is destined to coexist.

The disruptor must choose the most appropriate moment to enter the market or drastically change strategy because technologies develop at an exponential rate. The history of technological innovation is full of potential disruptors that could offer superior performance to rivals, but which were unable to identify the right moment to enter the market. Finally, how can disruption be managed? The first step is to predict the impact of a new disruptive business model on consumer behaviour. The second involves assessing the likely scope of the offer and estimating how many consumers could move to the new product. In the automotive industry, how many consumers would adopt a driverless car? The third step is to extend the analysis of the impact of a disruptive business model to related sectors; for example, in the automotive industry the analysis needs to be extended to the supply chain and the channels of distribution.

In the 2000s, traditional carmakers made an important initial reaction to the advances in the digital economy, starting to prepare themselves for their imminent future as "transportation solutions providers", and no longer merely vehicle manufacturers.

A new breaking point in the marketing of automotive companies was approaching, driven by new digital technologies that enormously expand the reach, speed, convenience and efficiency of platforms. The system of relationships rendered possible by the platforms changed and damaged the traditional value chain in the automotive industry. In the more advanced stages of the digital age, the advantages generated by technological advances did not in fact affect the supply side, but rather the demand side, and gave rise to significant demand economies of scale and network effects.

These new technologies facilitated communication and the exchange of data between participants in the network. The more a company attracted new participants to the platform (owners, providers, producers and consumers), the greater the network became and the more transactions between demand and supply increased. The capacity of certain companies to take advantage of the development of the platforms soon posed a threat for incumbents in the automotive industry.

New competitors, free from previous obstacles, quickly entered the market, proposing a new kind of "crowd-based public-private partnership", including Uber, Lyft and BlaBlaCar. The emergence of Tesla is also a clear example of a threat for incumbents.

In recent years, digital marketing in the automotive industry has made significant advances but carmakers were slow to respond to these advances in technologies. The research analyses the reasons.

In the second part of the 2000s, digital transformation is rewriting marketing rules and new strategies are arising. Nowadays, car clients are very different from those of the recent past. Their behaviour when seeking data and purchasing has changed quickly and they make ever greater use of digital media.

The amount of data available to manufacturers has also increased and the problem of how to quickly select those that are really useful therefore arises. It is important for carmakers to understand their audience and its behaviour during the purchasing process. Given that potential clients can find a huge amount of content online, they make many decisions concerning the purchasing process before visiting a dealership. The marketing content they can find online has a significant impact on these decisions. Moreover, it is not enough for car manufacturers to concentrate on their current customer target; they also need to know about those who are consuming information and dealing with brands digitally.

The development of digital technologies also marked the need to move and extend the "7Ps" of marketing (product, place, price, promotion, and also people, process and physical evidence). Manufacturers need to create value by combining tangible and intangible assets in ways and with extensions that had no precedents. New technologies made it possible to change the core product and extend it. From the early 2000s, car companies started to redefine their value propositions. They moved towards "selling personal transportation solutions" rather than "just selling cars". Customers now participated in creating products through the Web. In the pricing field, transparency increased, and downward pressure on prices grew. Regarding place, the relationship between customers and dealers is difficult to replace. Thus, given that many car buyers prefer to search for vehicles online and make their decisions before they even consider visiting a dealership, manufacturers often turn to augmented reality and virtual reality to present the features of their products to potential customers. In terms of promotions, the digital age opened up new channels of communication. Advertising, promotion, publicity, public relations and most other aspects of corporate communications are in part outdated concepts. In the digital age, customers have access to a large amount of information about products and "power shifts towards them". They have information not only from the manufacturer or intermediaries, but also from other consumers among themselves via the Web. What the customer notes and appreciates depends ever less on what the company communicates about itself.

On the eve of the 2020s, the concept of "digital transformation" is discussed increasingly frequently in the car industry, bringing a profound change in the competition in all sectors of the economy. Analysing the current situation regarding specific trends in the automotive industry, the aim is to offer a prediction as to the evolution of marketing strategies as we move towards the 2030s.

Three forces, acting together, created the new digital world. The first, in order of time, was the exponential growth and ever-lower costs of computer power. The second force was the value of networks, which has grown as they have grown in size, and the third is that more data have been transmitted at an ever-lower cost using cloud computing technology. Faced with technological change occurring at an exponential rate, companies are much slower. Filling or reducing the gap is the main challenge faced by management.

Furthermore, in the automotive industry four innovative trends merit particular attention: mobility services instead of vehicle ownership; increasing demand for connected services; autonomous driving; and electromobility (EV).

In major European cities, in cities in the USA, and in the megacities of other nations, the impact of car traffic on the environment and time wasted as a result of traffic congestion reduce the desire for car ownership, especially among young people, and the demand for mobility services increases, in turn, facilitated by technological advances. The mobility services offer is dominated by Uber, Lyft, Zipcar, BlaBlaCar and Didi, with similar user-friendly, low-cost services. Some traditional manufacturers offer mobility services, often in cooperation with partners, while others have created a distinct organisational division. It is plausible to predict that the trend towards mobility services will have a negative impact on the sales volumes of car manufacturers as it will reduce brand loyalty.

The second trend is the demand for cars equipped with connectivity features. Connectivity facilitates the use of mobility services and therefore expands the demand for them, giving manufacturers greater possibilities for new business models. The trend towards greater connectivity leads to collaboration between multiple operators, because the consumer wants continuity between one service and another, and above all fast response times.

Autonomous driving is an interesting subject, a trend that will take place in evolutionary steps. In the first three stages, control of the vehicle is in the hands of the driver, assisted by autonomous systems. These stages are a powerful marketing tool because drivers see them as elements that offer greater driving safety and are prepared to pay for them. The evolution from high automation to complete automation is difficult to predict in terms of both time and demand, but almost all manufacturers are investing significant amounts on this. This study has endeavoured to provide an explanation for this risky investment decision.

The last innovative trend arises due to traffic density, increasing pollution and noise. It concerns the demand for electric vehicles in large areas of the world. A thorough, comprehensive analysis and discussion of the advantages and disadvantages of the electric car is developed in this study. Manufacturers are investing in lowering the CO_2 emissions of their fleets, to meet the demands of governments and some customers, but the transition from traditional cars to electric ones (if it happens) will be slow: What strategies could be adopted in this period of transition?

In this time of digital transformation, there are no formulas to define a marketing strategy, but each company should analyse three specific issues to identify unique solutions. Consumers' desires are increasing to such an extent that they might be described as "unreasonable expectations". These consumers, accustomed to automatic responses, expect the same speed at all times and from whoever is offering the products. If they want a quote for a built-to-order car, they expect fast responses even for the most personalised, specific requests.

Responding to clients is becoming increasingly challenging, especially if waves of innovation intervene. On-board instruments have been improved: voice-activated functions, artificial intelligence, and augmented reality. The problem lies in understanding, during the design stage, which technology brings advantages that the client can appreciate in concrete terms. The second issue to be analysed is the competitive environment, which has been redefined by the digital revolution. Competition occurs less within individual industries and more between different industries (Google, Apple, Amazon, Uber, Lyft, Didi). Moreover, some carmakers compete among themselves in certain areas, but are partners in others. The third issue concerns understanding how digital marketing fits into the company's business model and how and what changes are needed in the processes. Big data must be managed; the nature of competitive advantages can change; new value propositions need to be offered; and the need for new organisational structures and culture increases.

The final part of the last chapter underlines the high probability that within ten years, demand, the structure of competition and the players involved in the automotive industry will be very different from the situation today. To attempt to predict where marketing strategies are headed, it is important to remember that we could be faced with a radical innovation, rather than an incremental one as occurred for a long time in the past.

Under the pressure of digital transformation, established carmakers must make broad adjustments to their marketing policies and strategies. They must defend traditional business segments and open new ones, and combine traditional and new business models, old and new technologies, especially because they are present in many geographical areas. Definitive dominant standards of product and process have not yet emerged, government policies are not clear and unambiguous, the value of the brand is changing, and new positioning strategies need to be planned. Costs and uncertainty are growing. The digital age is calling into question the milestones of an entire industry.

Can we reasonably believe that "A new machine will change the world again"? The conclusions drawn in this book propose an interpretation of the near future in the automotive industry.

At the end of this introduction and of the work I have conducted over many years, I wish to thank the people who made this research possible: the top managers in the car industry who have transmitted their knowledge to me with generosity and transparency, opening the doors of their businesses; the managers from different industries who have been available for interviews; the people who have participated in the surveys; those who read the drafts and gave me their opinions; and the university professors who taught me the value of knowledge and curiosity.

Turin, Italy   Elena Candelo
Professor of Strategic management and Marketing
University of Turin
elena.candelo@unito.it

Contents

Part IV   The Digital Age: The Changing Face of Marketing

Part I
The Mechanical Age: From the Early Years to the 1950s

Chapter 1
The Pioneers: Racing as the Main Sales Promotion Tool

Abstract Towards the end of the nineteenth century, the Europeans were the first to design and construct motor vehicles and sell the first cars, but it was in the U.S. that mass production was developed as the main constituent of mass marketing. Indeed, at the beginning of the following century, the U.S. became the foremost manufacturer, primarily thanks to the work of Ford, who introduced the Model T into the market in 1908. In both Europe and the U.S., the first cars were sold to rich individuals, who were passionate about innovation and set on standing out and showing off their status. The vehicles were built at the specific request of individual buyers, who generally purchased the chassis from one manufacturer and the body from another. Since the first vehicles introduced into the market often stalled, the sturdiest and most reliable constructions were primarily identified through endurance racing. One cannot talk about marketing in the modern sense of the term at that stage. Many important characteristics of today's car industry date back to those early years. First and foremost, it was then that the dominant car design and the importance of clusters came to light.

Towards the end of the 19th century, the Europeans were the first to design and construct a motor vehicle and sell the first cars, but it was in the U.S. that mass production as the main constituent for mass marketing was developed. At the beginning of the following century, the U.S. in fact became the leading manufacturer, primarily thanks to Ford, who introduced the Model T onto the market in 1908. As is always the case in the introduction phase of a new product in the market, the marketing strategy should have relied on heavy advertising and sales promotions to generate product awareness and stimulate sales. However, races, either of endurance or speed, were the main tools used to promote sales.

1 Invented in Europe

In the last two decades of the 19th century, in Germany, Nicolaus Otto, Carl Benz and Gottlieb Daimler, working in proximity yet separately, developed the four-stroke

© Springer Nature Switzerland AG 2019   3
E. Candelo, *Marketing Innovations in the Automotive Industry*, International Series in Advanced Management Studies, https://doi.org/10.1007/978-3-030-15999-3_1

gasoline engine and are considered as the inventors of the modern automobile. In 1877, Otto had produced a four-stroke engine, an idea that later formed the basis of the gasoline-powered modern engine. In 1866, perfecting Otto's invention, Benz patented the Patent Motorwagen, a single-cylinder engine adapted to a tricycle. In the years immediately following, Daimler and Maybach, unaware of Benz's invention, further fine-tuned Otto's four-stroke engine and installed it under the seat of a four-wheel engine.

In 1900, France was the leading producer of engine-driven four-cylinder vehicles. Many small entrepreneurs, attracted by the new invention, produced slow, unreliable automobiles at high costs. Only the names of a few companies are still known to this day, such as Blériot, Panhard and Levassor, and Hispano Suiza. Some of these, such as Peugeot, were better known as bicycle manufacturers (Lottman 2003). Panhard and Levassor and Peugeot produced engines licensed by Daimler and Maybach. They had picked up on the potential of the Germans' inventions, but managed to go one step further. Some of the innovations attributed to Panhard still underpin the standard configuration of the automobile today, such as the use of a four-cylinder engine, the engine installed on the front of the vehicle rather than under the seat (as the pioneers had done), "steering the vehicle with a wheel instead of a tiller, attaching a transmission with gears to the rear wheels, and engaging a clutch to change gears" (Rubenstein 2014).

The first cars were artisanal products, constructed individually by hand. Each different, the components with which the same manufacturer produced one car after the next also often differed. Cars were almost invariably built on a chassis, which transported all the components, and held the engine, gearbox, axle shafts, and wheels. The owner would buy the chassis and then a body from a specialist coachbuilder. Often, after some time, they replaced this with a more modern body; this practice was fairly widespread in Great Britain (Nieuwenhuis and Wells 2015). After the early years, some suppliers were soon able to offer engines, gearshifts, axle shafts and other components, allowing for a first standardisation that anticipated mass production. This was particularly true in France.

Speed was the greatest attraction. The experience of being faster than the mode of transport of the time, horses, had come far earlier with railways (1840), but the first people to drive a car showed everyone the appeal of driving an open-air vehicle wherever they wanted and at speed. Cars evoked fantasies and dreams. They summoned up crowds of curious bystanders as they passed by. They also encouraged dangerous behaviour and folly on the roads, but manufacturers and dealers went on selling (McCarthy 2007; Epstein 1928; Rae 1965).

Races were the tool use to promote, and thus market, automobiles. Endurance and speed races drew in crowds and provided a test for performance in terms of speed and endurance. Since the first automobiles often broke down, endurance races singled out the most robust and reliable models and showed the few people that could afford to buy them which were the best. Races provided enormous publicity. Winning meant sales, because enthusiasts of the new product wanted to present themselves as adventurous sportsmen, as well as having a symbol of their wealth and superior lifestyle. In the first years of the new century, European supremacy was attested to by the fact that William Vanderbilt, a member of one of the richest American families at the time, and many others used to buy cars in Europe to compete in races in the U.S.

2 The U.S. Leads the Way

In the U.S., as in Europe, the first automobiles were unreliable, expensive, constructed based on the requests of the few clients that could afford the high prices (customized, in today's language). They were driven by rich people that loved adventure, who were often prompted only by the desire to show themselves off as pioneers of new technology. Transport by horse and the transportation of excrement collected on the streets were still growing technologies.

The leading companies of the time, which could have invested some of their financial resources, ignored the budding car industry because they saw embracing a new technology not connected with their current business as risky. Their place was taken by small entrepreneurs, enthusiasts of new technologies, and mechanical experts. Most came from industries related to the mode of transport of the time: the construction of bicycles and coaches. Many small workshops built parts and accessories that were then assembled by dozens of car manufacturers, all of which suffered financial problems due to their small size and the difficulty of ensuring continuity for productions in the embryonic stage in the industry.

The new invention from Europe was driven by a period of strong economic growth, as well as in various other fields. The early years of the new century in fact represented a period of extraordinary technological and cultural growth in the U.S. However, the mass market was still far from being developed. "In the year 1900, few in the United States envisioned a mass market for the gasoline-powered road car" (Farber 2002).

As Farber recalls in *Sloan Rules*, the English language was reorganised to create a vocabulary that could describe the performance of the new product, borrowing a lot from French, with words such as chassis, limousine, garage, chauffeur, and automobile (Farber 2002).

Racing as a tool of sales promotion. In the U.S., as in Europe, car races were the main form of product promotion. Races between cities and uphill quickly became very popular. Henry Ford knew how to make the most of this. The first race in the U.S. was held at the beginning of the century near Lake Michigan. It took place during a snowstorm. In an hour, the winner covered on average less than seven miles. The race was followed by many others, all of which generated great enthusiasm and publicity (Landes 2008).

Henry Ford was a successful racing driver. He built the cars he used to race on his own, and won many competitions. A large part of his success was thanks to the reliability of his cars. He built light but resistant vehicles, unlike many of the cars of the time which were built like vehicles not drawn by horses, heavy and unsuitable for travelling at speed along uneven roads often full of holes. From the earliest experiences, Ford thought about how cars could be used by farmers, the world in which he had been born and lived, instead of those who wanted a luxury vehicle (Landes 2008).

3 Dominant Design and Clusters

Many important characteristics of today's automotive industry originate in those early years. Above all, this was when the dominant design of the automobile and the importance of clusters emerged.

Dominant design. A typical phenomenon emerged in the new industries of assembled products, as described by Utterback (1994).[1]

A pioneer launches an idea and introduces the first configuration of a new product. In this first stage, both the market and the industry are in a fluid state of development. Everyone, manufacturers and clients, is learning. For as long as the financial and technical barriers remain low, many new companies enter the market, attracted by the innovation. Existing and new companies continually perfect the original product and gradually converge around common solutions. This continues until, at a certain point, "some centre of gravity eventually forms in the shape of 'dominant product design'. Once the dominant design emerges, the basis of competition changes radically" (Utterback 1994). After a dominant design has been permanently established, the number of firms in the industry declines until it becomes fairly stable. Successful firms are often among the ones that entered the industry at the beginning.

Utterback (1994) underlines that, during this embryonic stage of development in the industry, the innovation of a new product creates a position of temporary monopoly for one or several companies, and that this is in line with Schumpeter's concept of the "*creative destruction*" model and the successive developments in the study of the economy of innovation.

The story of the early years of the automotive industry is filled with dominant design experiences. The winning configuration of the automobile was: an internal combustion engine; four wheels with tyres; a steel structure; and in-vehicle electric systems for activating the equipment.

From the outset, the engine was most important element. In the early years of the last century, vehicles powered by steam engines, electric engines and gasoline engines were in circulation. Those with steam engines were easier to construct than those that burned gasoline, also because they used a technology already established in railway transport and agricultural machinery. They quickly declined as engines for road vehicles, however, because they were too heavy and took time to be pressurised, almost half an hour in fact. Those powered by electric engines were common only in

---

[1] Utterback (1994) defines a dominant design as follows: "A dominant design in a product class is the one that wins the allegiances of the marketplace, the one that competitors and innovators must adhere to if they hope to command significant market following".

urban areas, as they did not have enough power to travel along very badly connected rural roads, which often had a muddy surface. Moreover, the batteries were bulky and had to be charged up every 20–30 km. The internal combustion engine (ICE) emerged thanks to its superiority in resolving build-up and in-vehicle engine transport (petrol and diesel) problems, and due to the limitations of alternative technologies. Progress in terms of vehicle speed gradually led to improvements in suspensions, steering-wheels, and the use of tyres.

Clusters. The second characterised that emerged in the early years proved that inventions rarely occur in isolation from the industrial context. The precursors—Benz, Daimler, and Maybach—had worked separately, but within the defined geographical area of Germany. Almost half the cars produced in the U.S. at the beginning of the century were assembled in south Michigan. This area attracted the most successful manufacturers because it was home to mechanics and technicians specialising in three essential components: an engine to propel the vehicle; a drive train to convert the power of the engine into movement; and a chassis resistant enough to transport both the passengers and the engine. All these problems could be resolved by companies operating in a single area: the south of Michigan (Rubenstein 2014).

This phenomenon, wherein firms from the same industry come together to produce in a limited geographical area, is called "clustering". By working closely together, small firms can build the economies of scale reserved for larger ones. Small firms gain a great advantage from having easy access to the know-how of a skilled workforce and to parts and components offered by suppliers, as they can enjoy the scale of a large firm without the risk of making huge investments or the hurdles to quickly adapting to changes in demand.

After WWII, the success of Toyota's production system was assisted, to some extent, by the fact that Toyota's main factory in Japan was situated in Toyota City, near Nagoya, where there is a classic "cluster" of suppliers and manufacturers in the car industry. With its main suppliers just a few miles away, Toyota can choose parts and components on demand and be sure they will arrive at its assembly line exactly when they are needed (just in time, JIT).

References

Epstein R (1928) The automobile industry: its economic and commercial development. A. W. Shaw Company, Chicago

Farber D (2002) Sloan rules: Alfred Sloan and the triumph of general motors. The University Chicago Press

Landes D (2008) Dynasties. Fortune and misfortune in the world's great family businesses. Penguin Random House

Lottman H (2003) Michelin men: driving an empire. I. B. Tauris, London

McCarthy T (2007) Auto mania: cars, consumers and the environment. Yale University Press

Nieuwenhuis P, Wells P (2015) The global automotive industry. Wiley, London

Rae J (1965) The American automobile: a brief history. University Chicago Press

Rubenstein J (2014) A profile of the automobile and motor vehicle industry. Business Expert Press, New York
Utterback J (1994) Mastering the dynamic of innovation. Harvard Business School Press

Chapter 2
The First Paradigm: Mass Production and Mass Marketing

Abstract The move towards mass marketing occurred in the U.S. in the first decade of the previous century. Henry Ford understood that the potential market for a means of transport affordable by the masses was enormous. Prices could be kept down by producing large volumes of identical cars, to take advantage of economies of scale and reduce product costs per unit. Mass production and mass marketing were born. With regard to marketing, Ford assumed that what the customer wanted was an affordable price. He recognised that this approach could lead to a downward spiral of lower costs allowing for lower prices, with the latter leading to greater volumes, which, in turn, would allow for even lower costs. Renown was guaranteed through the popular appeal of innovation in terms of both the production process (assembly line) and the product. However, Henry Ford ignored the drivers of change. He failed to understand that the rules for success were changing and to adapt his marketing strategies to the changes in consumers' buying behaviour. Consumers were waiting for up-to-date models. He was so convinced of his choices that he persisted in using outdated production and marketing strategies. While other carmakers placed their bets on increasing wages among the population and on a growing interest in cars, Ford restricted his offering to a product for the masses sold at a low price. With this marketing strategy, he could not compete.

Towards the end of the first decade of the last century, Ford's innovations in production and marketing drastically changed the structure and dynamic of the automotive industry. Using the modern language coined by Christensen, Henry Ford was a true "disruptor": "The Model T… created the first mass wave of disruptive growth in automobiles" (Christensen and Raynor 2003).[1]

In another of his books, *The Innovator's Dilemma*, Christensen drew a distinction between "sustaining technology", developments that help firms make marginal improvements to what they are doing, and "disruptive technology", unexpected breakthroughs that force firms to rethink their mission as existing products and markets can be overturned. In his follow-up book, *The Innovator's Solution*, Christensen

---

[1] In *The Innovator's Solution*, Christensen and Raynor (2003) attributed the role of 'disruptor' of the industry to Henry Ford and the Model T. "Henry Ford's Model T was so inexpensive that he enabled a much larger population of people who historically could not afford cars to own one".

© Springer Nature Switzerland AG 2019

E. Candelo, *Marketing Innovations in the Automotive Industry*, International Series in Advanced Management Studies, https://doi.org/10.1007/978-3-030-15999-3_2

changed the term from "disruptive technology" to "disruptive innovation". One problem with "disruptive innovation", Christensen argued, is that it often arrives unno-
ticed, little by little and from unlikely directions. It rarely comes from large established organisations.

Tedlow (1996) assigned an even larger role to Henry Ford: "It was Ford first and
foremost, who had the vision of what a car should be. It was he who in the first
decade of the last century refused to go in the direction of low volume, high margin,
and high prices. He and his people embraced the challenge of mass production."

There is no doubt that Henry Ford was a "disruptor" of his time and a forerunner
of modern marketing management. Henry Ford did not use the term "marketing
concept" but was very adept at understanding its logic and chain of reasoning.

(1) Before starting production, he assessed marketing opportunities (demand for a
    simple and low-priced mode of transport), evaluated the potential demand for
    automobiles, and selected the right target of potential customers.
(2) Ford then developed the right product to reach the target. He identified the
    product characteristics desired by consumers (simple to use and repair) and
    offered a product that matched the needs and desires of the target market more
    effectively and at lower prices than his competitors.
(3) Having selected the right technology, he was able to offer an affordable price
    for the masses. He was not interested in segmenting the market. His objective
    was to find the best product and sell it to everyone (Tedlow 1996).
(4) He knew how to motivate his workers.
(5) He recognised the importance of distribution and of building close relationships
    with customers. After a first attempt, he was forced to abandon direct distribution
    as a means to sustain customer relationships and shifted to franchised dealers.

1   The Right Target

Henry Ford soon succeeded, because he had understood the market potential and
chosen the right target. He had listened to customers early on. He was aware of
what other pioneers had ignored. There was a strong, latent demand for vehicles.
Beforehand, cars had primarily been an object of luxury and a status symbol. Ford,
however, wanted to attract demand among anyone that needed a vehicle that was
easy to drive and affordable. He built the right product to offer these characteristics.

2   The Right Product

Ford focused his attention on designing a new car that was small, light, robust,
reliable, fast for the time and affordable by consumers with an average income.
Observing how a French racing car that had been smashed up during a competition

was built, Ford understood that he had to get hold of a new type of steel: vanadium steel. He thus succeeded in building an alloy that made it easier to configure the exterior of the vehicle and ensure the entire structure was sturdy enough to resist any unpleasant consequences and damage caused by rough roads. The Model N was created, followed by the Model T, the car that made Ford famous.

When the first cars to be tested began to go into circulation, dealers were apparently incredulous. When the first Model T was sold in 1908, orders from dealers came pouring in and the operating capacities of Ford's factory were quickly surpassed. Tin Lizzie, as the Model T was jokily known, became the most popular car, and not only in America. For eighteen years, Ford sold a product that only came in black, because simplifying meant reducing production costs and black paint took less time to dry than other colours (Landes 2008).

3   New Technologies

Ford began building cars on a large scale in 1910, when he opened a plant at Highland Park in Detroit. Before that, his workers were artisans with manual skills, mechanical experts who took around 33 h to build a Model T. In the years following the opening of the new plant, Ford drastically reduced the time required to assemble his products. Production was accelerated by introducing machines that cut down on labour. Instead of using specialised artisans to build costly cars, he used machines to assist a non-specialist workforce to produce cars in less time and at lower costs. Only 2% of his workers could be classified as mechanics or foremen. He had revolutionised the car industry by introducing mass production methods. The innovation of the moving assembly line, where the product came to the worker rather the worker to the product, was rapidly adopted by other industries and in other countries. The automobile had rapidly become a product for the masses.

4   Motivated People

Ford grasped the importance of motivation in work and of a sense of belonging to the company. In exchange for strict rules and the monotony of the work along the assembly line, he doubled salaries, bringing them up to five dollars for eight hours in the factory. In this, too, he was an innovator. Instead of paying employees based on the number of vehicles produced, the prevalent method in Detroit, he introduced a daily salary. Five dollars was two to three times more than other manufacturers paid. At that rate, it took a Ford worker around 80 days to earn enough to buy a car, much less than it takes nowadays to buy a mass market car.

Ford's new contract gave his employees more free time and money. Previously, they had had to work for hours and hours to buy food, a home, and clothes. The increase in pay improved Ford's reputation, as well as that of his products and the entire industry. The national newspapers all reported on the story and described the lines of people in front of the factory gates, hoping to be taken on. The decision to increase pay helped initiate a period of mass consumption in American society. The increase in purchasing power fed mass entertainment, the dissemination of the cinema, and the development of radio networks. This all contributed to the emergence of a new lifestyle of which the Model T was part.

5   Affordable Prices for the Masses

In marketing, Ford assumed that what customers wanted was affordable prices. He understood that this approach could lead to a downward spiral of lower costs resulting in lower prices, with those lower prices leading to greater volumes which, in turn, would allow for even lower costs. In 1908, Ford sold his first cars for $825 (Landes 2008). This price was no small sum for the time, but it was considerably lower than the average of other vehicles for sale, which hovered at around $2000. Years later, in 1923, the new Model T was sold for $295. The price went down every year. "The auto had become an affordable everyday necessity for millions of Americans" (Farber 2002). When Ford ceased producing the Model T in 1927, fifteen million units had been produced and sold. Ford had revolutionised the business.[2]

6   "Genius" in Marketing

Henry Ford's assembly line is widely considered to be the start of the industrial revolution, but Theodore Levitt, as expressed in *Innovation in Marketing*, believed differently. He wrote that, "Henry Ford's real genius was marketing". We believe, Levitt posits, that Ford was able to lower sales prices and therefore sell millions of cars because his invention of the assembly line reduced costs. In fact, however, he invented the assembly line because reducing costs and prices would sell millions of cars. Mass production was the result not the cause of his low prices (Levitt 1962).

7   Direct Distribution Was Abandoned

The first people to purchase cars ordered them directly from a manufacturer. When mass production began to become widespread, they could buy them in a bicycle

---

[2]Ford wrote that: "I will build a motor car for the great multitude... It will be constructed of the best materials, by the best man to be hired, after the simplest designs that modern engineering can devise. But it will be so low in price that no man making a good salary will be unable to own one..." (Chandler 1990).

shop, a hardware shop or in other retail outlets that had some affinity with mechanics. However, this solution did not meet the requirements either of consumers or of manufacturers. Such intermediaries rarely had the necessary knowledge to assist buyers before and after the sale. Moreover, the cars were difficult to drive and unreliable, and buyers rarely knew what they were really purchasing.

In the early years, Ford sought to work alongside independent dealers through his own outlets and thereby gain the benefit of direct contact with consumers. He was soon forced to give up, however. This was for two reasons: first and foremost, the employees at outlets subordinate to Ford displayed scant levels of motivation to sell compared with independent dealers; moreover, and this was a far more important fact, the rapid growth in demand for cars surpassed the capacity to create a sufficient number of outlets owned by Ford and to cover the territory adequately. In 1910, Ford's dealers numbered around one thousand. Ten years later, this figure had reached six thousand units. According to Nevis and Hill (1954), Ford "Never met any difficulty in selling all the cars he made". After the first two decades of the century, vehicles started to be sold exclusively through "franchised dealers" (Rubenstein 2014).[3]

8   Advertising and Sales Promotion

Price and the product's technical characteristics were the two key questions in advertising. In their advertising messages, newspapers belonging to the Ford company conveyed that the car was sold at a low price, easy to drive, and useful for various purposes in factories and for product deliveries. Ford wanted to offer something useful rather than something stylish. With the claim "Facts from Ford", an advertisement in 1912 stated that the price of the Model T Touring Car had gone down to $690 f.o.b in Detroit. It went into numerous, expansive illustrations of the product (no fewer than 21, plus four price variants), such as "The lightest weight 4-Cylinder motor car in the world", "The cheapest 4-Cylinder motor car in maintenance", "The simplest motor car in design" and "Remember that Ford motor cars are sold fully equipped" (Vaikin 2008).

As the historian David Lewis explained, Henry Ford had mixed feelings about advertising, admitting that it was "absolutely essential to introduce, good useful things", but "an economic waste" for products already on the market. Lewis estimated that Ford's expenditures in advertising were lower than those of any other major consumer firms in the 1910s and stated that from 1917 to 1923 he did not spend a penny advertising the Model T (Lewis 1976). This policy was made easier by two facts. Firstly, under the terms of their contracts, the company imposed advertising expenditures on dealers. In this way, the company benefited from an estimated three million dollars a year in advertising, a huge sum for the time. Secondly, Ford was able to spend so little on advertising thanks to the value of the product and the free publicity generated by his personal fame as CEO (Lewis 1976; Flink 1988).

---

[3]The quoted sentence and remarks come from Rubenstein (2014, pages 82, 84).

The use of advertising and sales promotions corresponded with the stages of the product life cycle we are familiar with nowadays (in the case of an entirely new product for the market). The life cycle includes four phases: introduction, growth, maturity, and decline. Each phase calls for a different marketing strategy. In the introduction phase, sales are slow and profit is non-existent, and the marketing strategy should therefore mainly rely on advertising and sales promotions to create product awareness and stimulate sales. That is what Henry Ford did.

9   Complacency

Drivers of change were ignored. In the early 1920s, the Model T sold millions of units and made up around 90% of the low-cost vehicles market. In America, however, an increasing number of people bought automobiles and new competitors offered innovative models in competition with Ford, in particular GM. Ford had understood the risk, but he was so convinced of his choices that he persisted with outmoded production and marketing strategies. While his competitors offered a pedal accelerator, he continued with a hand one. While GM offered a wide choice of colours, Ford mainly offered black. Ford's cars were cheap, but five or six other manufacturers offered prices below $1000 (Landes 2008).

The evolution of the business model from a family business (Ford) to a joint-stock company (General Motors) also influenced the evolution in marketing. While Ford was a great centraliser, to the point of becoming a despot in the last years, Sloan (GM's CEO) was accountable to owner-shareholders. While GM gambled on the increasing wages of the population and an ever-greater interest in cars, Ford limited his offering to a product for the masses sold at a low price. With this marketing strategy, he could not compete.

Ford failed to understand that the rules for success had changed. "By 1921, Ford's strategy of competition strictly on the basis of price was beginning to lose its power… The time of the great surge of first-time buyers was drawing to a close. The world of automobiles had changed" (Tedlow 1996).

He failed to adapt to the changes in consumers' buying behaviour and ignored his son Edsel's advice to terminate production of the aged Model T. His firm collapsed and its market share was halved (reduced from half to a quarter). Consumers expected up-to-date models. Before the end of the 1930s, GM had relegated Ford to second position (Landes 2008).

10   The Paradigm Concept

Ford's decline did not overshadow the advantages of mass production. The Model T marked the beginning of a new way of producing. The results obtained in the U.S. convinced managers, not only in the car industry, that was the route for success. It became the first paradigm of mass production and mass marketing, and was then also adopted in the more advanced European countries.

The paradigm concept, as applied to science in general, was introduced by the scientific historian Thomas Kuhn, who defined it as "an accepted pattern or model" that establishes a set of rules by which its practitioners view the world (Kuhn 1970). It is a perception shared, over an extended period, by at least the majority of researchers (managers in our case) concerning a way of viewing the world in a given historical period (Sundbo 1998). Applying the paradigm concept to business, Barker defined it as that which "tells you that there is a game, what the game is, and to play it successfully". Ford's principles of mass production had taught firms how to play the "game" with success (Barker 1988). The rules of the paradigm were then completed by another innovation owed to Edward Budd.

Ford had introduced innovations by organising certain production techniques at the time in a new way. As for the product, he started with the modular approach used by the foregoing craft builder: a chassis separate from the body, which was coach built. Today, mass production uses an all-steel vehicle underframe and a body built as a single rigid structure ("monocoque"). This technique, known as the all-steel welded body, was introduced by Edward Budd around the time that Ford introduced the moving assembly line and is considered by many to be an integral part of the first paradigm of mass production (Grayson 1978; Nieuwenhuis and Wells 2015).

In the modern version, the Budd-style steel body technology requires major investments in plants (presses, welding, painting) which, when made through economies of scale, translate into low product costs per unit. For this reason, some authors, including Nieuwenhuis and Wells (2015), argued that "Budd's innovations constitute the very basis for the economics of the industry".

Budd's technology had a major impact in Europe and the U.S. It helped reduce the weight of vehicles. Given that the U.S. was self-sufficient in terms of fuel supplies and energy costs were low, a culture of accepting low levels of efficiency in the use of energy began to spread among designers, manufacturers, and clients.

Conversely, in Europe none of the main car manufacturing companies were self-sufficient when it came to fuel supplies. Consequently, European governments set high taxes on consumptions from the outset, and thus on engine power, to protect their balance of payments since oil had to be imported. Budd's all-steel technology was introduced in Europe by Citroen in 1934. According to certain authors, this date marked the start of the development of the mass car market in Europe.

References

Barker J (1988) Discovering the future: the business of paradigms. ILI Press, St.Paul, MN
Chandler A Jr (1990) Scale and scope. Harvard University Press, MA
Christensen C, Raynor M (2003) The innovator's solution. Harvard Business School Press
Flink J (1988) The automobile age. MIT Press, Cambridge, MA
Grayson S (1978) The all-steel world of Edward Budd. Autom Q XVI(4):352–367
Kuhn T (1970) The structure of scientific revolutions. The University of Chicago Press

Landes D (2008) Dynasties. Fortune and misfortune in the world's great family businesses. Penguin
    Random House
Levitt T (1962) Innovation in marketing. McGraw-Hill, New York
Lewis D (1976) The public image of Henry Ford: an American folk hero and his company. Wayne
    State University Press, Detroit
Nevis A, Hill F (1954) Ford the times, the man, the company. Scribner's, New York
Nieuwenhuis P, Wells P (2015) The global automotive industry. Wiley, London
Rubenstein J (2014) A profile of the automobile and motor vehicle industry. Business Expert Press,
    New York
Sundbo J (1998) The theory of innovation: entrepreneur, technology and strategy. Edward Elgar,
    Cheltenham
Tedlow R (1996) New and improved: the story of mass marketing in America. Harvard Business
    School Press
Vaikin J (2008) Driving it home: 100 years of car advertising. Middlesex University Press

Chapter 3
The Metamorphosis of the Automotive Market

Abstract Towards the end of the 1920s, a second turning point in the evolution of car marketing came about. General Motors ("GM") overtook Ford in terms of sales volumes by offering more attractive vehicles. Contrary to Ford's approach, GM concentrated more on the styles, colours, and external appearance of its vehicles. GM's marketing strategy was geared at segmentation: "a car for every purse and purpose". The strategy was based on the class divisions that were coming to shape American society. Economic development favoured social climbing. As the average income increased, more people climbed the "ladder of consumption", acquired a new status and had more time at their disposal to enjoy a new lifestyle. For the different social classes, GM offered different products at different prices, seeking to attract potential clients from the various segments through different advertisements. When he became President of GM in 1923, Alfred Sloan established a new relationship within management between marketing and engineering, and marketing and production. In just a few words, he rewrote the rules of marketing. He understood the importance of changing consumer expectations. In this sense, he too was a "disruptor". The motto "A car for every purse and every purpose" not only established the principles of segmentation, but also signalled the abandonment of the rule of considering a product merely in physical terms. GM devised its products not simply as means of transportation, but as objects, as tools for attracting and eliciting positive perceptions among potential customers.

"During the first part of the 1920s [...] certain changes took place in the nature of the automobile market which transformed it into something different from what it had been all the years up to that time", wrote Alfred Sloan in *My Years at General Motors*. This was a fortunate event, Sloan added, because as the challenger of Ford's position, which had been dominant up to that point, GM benefitted from the change. Since it had not based its growth on the "old" business model, the change brought GM new opportunities and a great advantage.

In the U.S., those years in fact marked a discontinuity with the past for various industrial sectors. The development of mass markets initiated a chain of consequences. It brought high investments in production and marketing to make the most of economies of scale and scope. In certain capital-intensive industries, a new type of

© Springer Nature Switzerland AG 2019
E. Candelo, *Marketing Innovations in the Automotive Industry*, International Series in Advanced Management Studies, https://doi.org/10.1007/978-3-030-15999-3_3

company emerged, which began to modify the competitive environment inexorably. The chain of consequences was at the base of what Chandler defined as the "visible hand" of professional managers (Chandler 1963, 1990).

1   GM: A New Marketing Concept

While most car companies, both in Europe and in the U.S., were established as family companies, and sought autonomy from suppliers through vertical integration once they had passed the embryonic stage, GM had a different origin and development.

William Durant, a shrewd, unscrupulous financier, used entirely borrowed funds to purchase a series of struggling firms in the nascent car industry, and in 1908, the year the Model T was introduced, he grouped them together under the name of General Motors. He believed that diversification could minimise risk. Some specialised in the production of parts and components, while the largest firms assembled automobiles. Since they each had their own origins, independently of the others, and since, as a matter of principle, Durant had granted each one nearly total autonomy in choosing what to produce and in setting prices, there were numerous overlaps in terms of products and markets. This all made the new company vulnerable since it scattered resources while it had to face up to new competitors attracted by the development of the industry and a rapidly developing Ford Motor. Alarmed by Durant's financial difficulties, the bankers ousted him.

Durant did not give up, however. He resumed purchasing insolvent part suppliers and assemblers and in 1916 regained control of GM. Nonetheless, as had already occurred, the excessive debt he contracted from repurchasing GM whittled down the resources needed to equip the group with a strategy in keeping with the times.

Once again, the history of the automotive industry was marked by changes in the economic environment. During WWI, there had been an increase in spending by the American government to purchase war materials, along with a rise in purchases of all types of food products and merchandise by European countries at war. This stimulated the entire American economy. 1919 was a record year for sales both by GM (which occupied 20% of the market at the time) and by Ford (50% of the market).

Towards the mid-1920s, the situation had changed again. The fall in European demand brought down the prices of agricultural products and American industrial production reduced on account of the conversion of many plants from the production of weapons to productions for civil purposes. The U.S. entered a period of recession. There was a drastic drop in sales of new cars and GM's financial situation quickly worsened.[1]

Price war. As aggressive as ever on the sales front, Ford reduced the prices of the Model T by around 20–30% and forced his dealers to absorb part of this price

---

[1] This period of American economic history and the crisis of GM are described in detail in Chandler (1963).

reduction. GM deteriorated further and was placed under pressure by bankers alarmed by the slump in the value of its shares. In November 1920, Durant gave up the chairmanship of GM.

The bankers' attention was centred on Pierre du Pont, who they wished to take over the chairmanship to protect the substantial investments they had made in the company. Du Pont accepted, but immediately declared that he did not wish to assume a position of control in terms of operational management. Although he was only fifty, he had held great responsibilities in the past during his successful management of a major chemical company, which bore his name, with which he had built up incredible amounts of money, particularly through the sale of arms to European countries during WWI.

"Organization study". Alfred Sloan, the owner of one of the many suppliers of parts and components acquired by Durant, held a secondary position in GM at the time. Nonetheless, he decided to directly contact the new chairman of GM, Pierre du Pont. In a letter, he explained that he had prepared a reorganisation plan, which had been approved by many of the senior executives of the company (Farber 2002). Du Pont expressed an interest in the plan. Sloan immediately sent him a document entitled "Organization study", considered to be a demonstration of genius among academicians and practitioners of management. "Organization study" set out a decentralised organisational structure based on several divisions, each with its own market and responsibilities, freedom of activity, and return on investment (ROI) targets.

From that point on, and in the following decades, GM used ROI to monitor the results of its divisions, as well as those of its competitors.

Each division was asked to demonstrate, on a monthly basis, that its ROI had met the target of 20% set for all of GM. Division manages were free to determine how to meet this objective. The "organization study" was considered as a sort of invention of major, decentralised modern companies.

Pierre du Pont wished to appoint Sloan as his first assistant; they had both studied engineering at Massachusetts Institute of Technology (MIT), and they were only five years apart in age. Above all, he appreciated the plan developed by Sloan, which granted a new organisation to GM based on two principles. The first was the need to maintain the business independence of those in charge of the various divisions, while the second was the need to create an efficient centre, later termed 'corporate', to guide and monitor the various divisions (Farber 2002). The new plan was accepted by GM's Board of Directors in the last days of 1920, and Sloan was entrusted with its implementation.

With the new organisation, "federal decentralization" as Sloan called it, each division had its own market segment, target and marketing strategy. Each division was run as a company within a company. Sloan said that the company was "co-ordinated in policy and decentralized in administration". This innovation marked an important step in the evolution of marketing strategies.

1921 was a difficult year for GM. The number of first-time buyers was low and looked set to decrease further. Moreover, Ford had a considerable advantage since the prices of his vehicles were much lower than those of his competitors, including GM. GM's management had no choice but to focus on the richest individuals, but they represented a very limited market and demand was also competed for by many competitors who, also as a result of the difficulties they were experiencing, were willing to accept low prices just to survive. Between 1921 and 1929, the American economic situation improved, as did that of the automotive industry, with sales rising from 1.7 to 4.3 million vehicles. Thanks to its new organisational structure and marketing strategies, GM managed to get a head start over all its rivals.

"Federal decentralization" is said to have taken only a month to set up, but its consequences were long-lasting and its results both enduring and impressive. Within six years, the company had gone from lagging behind in the industry to being the market leader, with a turnover of $1.5 million and a share price that had almost quintupled. Alfred Chandler provided a masterly description of the details and logic of the new organisational idea in *Strategy and Structure: Chapters in the History of the American Industrial Enterprise.*

When Ford was forced to give up on the production of the Model T in 1927, GM had obtained 40% of the American market. In the years immediately following, the solidity and efficiency of the marketing strategies were once again demonstrated when sales in the market decreased due to the Great Depression, and GM managed to acquire an increasing quota of a market with falling demand.

2   The Watershed: Towards the "Mass Class Market"

What were the factors of GM's success? What role did the new marketing strategies play? In his book *My Years with General Motors*, Sloan (1964) distinguishes three periods in the history of the automobile in America at the beginning of the last century: the class market; the mass market; and the mass class market. The class market corresponded to the period before 1908, when automobiles were purchased by rich individuals that could afford to pay a lot for an unreliable product. Then came the mass market, between 1908 and the mid-1920s, a period governed by the concept introduced by Henry Ford of a vehicle as a basic mode of transport sold at a low price, which gave many Americans their first automobile. This was followed by the period of the mass class market, wherein cars on the mass market improved continuously and were offered with great variety. Sloan identified the third period with "his" new GM. "This last I think I may correctly identify as the General Motors concept". The three periods had in common the positive effects of the long-expanding American economy and continuous population growth.

Sloan then listed four elements introduced by GM which, more than others, helped transform the market and create the watershed, the line that separated the *mass class* market from the previous period (the mass market) and helped mark the end of Ford's supremacy: (1) "instalment selling"; (2) the "used-car-trade-in"; (3) the "closed body"; and (4) the "annual model".

- **Instalment selling**. The sale of cars through borrowing was not new, but it spread considerably, to the point that in 1925 it made up 65% of car sales. Borrowing enabled those with a good salary to also purchase very expensive products such as automobiles. GM's management was convinced that with the constant increase in the average income of Americans, many would aspire to purchase quality cars and would be willing to pay higher prices. *Instalment selling* should have stimulated this trend.
- **The used-car trade-in**. During the 1920s, GM introduced the *trade-in* as a down payment on a new car. Although the trade-in was initially conceived as a means of getting old vehicles off the road, activity for it quickly became clear that sales of second-hand cars could become an additional, profitable dealers (Flink 1988). When car buyers, often first car buyers, handed in their old car as a down payment, a new phase in trading business between dealers and clients began. The new trend also influenced the characteristics of production inasmuch as dealers began to sell to people who already had a car, which they used as a first payment on the new car. Comparing the production data with the cars in circulation, and bearing in mind that cars were most likely traded in two or three times before being scrapped, Sloan estimated that between 1919 and 1929 there was a rising curve of used-car trade-ins. Rising incomes helped many Americans supported by instalment selling and used-car trade-ins to create "the demand not for basic transportation, but for progress in new cars, for comfort, convenience, power and style" (Sloan 1964).
- **The closed body**. Before WWI, the production of closed-body cars was a rarity desired by a limited number of customers. In less than ten years, however, between 1919 and 1927 closed-body cars went from occupying 10–85% of the American market. Ford was unable to benefit from this trend because the production technology of the Model T did not allow him to do so. The Model T was in fact designed as an open car. Its light chassis could not support the weight of closed-body automobiles. When the demand for closed-body cars went up, in a few years the Model T became an obsolete product.
- **The annual model**. For several years, GM avoided using the expression "annual model", even if each year the policy of upgrading led to changes to the products. It was not until the 1930s that this was declared explicitly. GM thus introduced the concept of planned obsolescence. Every year, models changed and were designed not to compete with each other.

  This practice also influenced strategies in other industrial sectors, helping to create the modern culture of consumption. It was deliberately ignored by Henry Ford, however. In his memoirs, Sloan wrote that he had always thought that the concept of continuous improvement was extraneous to the product policy of his rival and that when, in 1928, Ford stopped producing the Model A (which had replaced the Model T), although it was a good vehicle in its category, "it seems to me [it] was another expression of his concept of a static model utility car".

3 "Car for Every Purse and Purpose"

Ford had designed the type of automobile suited for the masses of the 1910s, and in part that of the 1920s. Towards the end of the latter decade, however, GM managed to sell to clients of every social class, from workers to the nouveaux riches and aristocrats. In those years, GM developed the concept of the product range, enabling clients to gradually "trade up" from the basic Chevrolet model, to the Oackland (later called the Pontiac), Buick and Oldsmobile, before finally arriving at the Cadillac. Unlike what Ford had done, GM concentrated more on the styles, colours, and external appearance of the vehicle (Flink 1988).

Different advertisements for different products. During the 1920s, GM overtook Ford in terms of sales volumes. It offered vehicles with a more attractive style. Sloan gave his segmentation strategy the objective "a car for every purse and purpose", an expression that became the advertising claim, but which also signalled a new stage in marketing.

The new marketing strategy was based on the distinction into classes that was coming to shape American society. Economic development favoured social climbing. As the average income increased, more people rose up the "ladder of consumption" and changed their status and expectations. For the different social classes, GM offered different products at different prices, and sought to attract potential clients from different segments with different advertisements. He proposed high performance with the Pontiac, cutting-edge technology with the Oldsmobile, reliability with the Buick, and a symbol of wealth and power for Cadillac owners.[2]

Marketability. Long before considering the marketing revolution of the early 1950s, Sloan understood the importance of changing consumer expectations. In this sense, he too was a "disruptor". When he became chairman of GM in 1923, Alfred Sloan established a new relationship between marketing and engineering, and between marketing and production within management. In a few words, he rewrote the rules of marketing.

"A car for every purse and purpose" not only established the principles of segmentation, but also signalled the abandonment of the rule of considering a product simply as a physical product. In assigning different market targets to the various divisions of GM, Sloan began with the principle that the objective of product design processes was marketability: "Products were considered means of behavioral control, not ends in themselves" (Kotler and Armstrong 1999).

While Henry Ford moved from the concept that people could buy his cars in any colour they preferred, provided it was black, GM devised its products not simply as

---

[2]"GM's marketing structure worked because it both reflected and shaped the American class structure... GM's strategy brilliantly reflected the realities and aspirations of the American family, and corporate advertisements listed the products by price. Chevrolet was the car for the masses; Cadillac, for the aristocrats; and the other three cars (Buick, Pontiac and Oldsmobile), for the growing middle class in between" (Rubenstein 2001).

modes of transport, but as objects, tools for attracting and for eliciting favourable perceptions in potential buyers.

4 The Second Paradigm: Multi-Product Mass Marketing

The Henry Ford (and Budd all-steel body) paradigm was thus followed by a more marketing-oriented paradigm, which was more complex and more suited to the evolution in demand conceived by Sloan, the CEO of GM. While the Ford-Budd paradigm was principally based on mass production, that of Sloan also comprised the production and marketing of a plurality of products, in addition to mass production. It proved more suited to a market geared at new structures, where clients wanted a varied offer. The winning marketing strategy led GM to dominate the market, to such an extent that the authorities regulating competition were led to intervene.

Between 1950 and 1955, GM's average ROI was around an impressive 25%. GM was so large and powerful that the Justice Department wanted to break it up. Since its market quota was rising, in 1956 the government warned that it would have to take "extreme action" unless GM slowed down its push towards monopoly. Washington suggested that GM could better achieve these objectives by selling one or two of its divisions. As well as dominating in the car sphere, GM produced 43% of the industrial vehicles in the U.S., constructed 60% of all diesel engines sold worldwide, and was the largest producer of refrigerators with the brand Frigidaire. "The attention of Government was well deserved. Besides being the biggest car company, GM was also the most proficient" (Taylor III 2010).

References

Chandler A Jr (1963) Strategy and structure. MIT Press, Cambridge, MA (reprinted by BeardBooks)
Chandler A Jr (1990) Scale and scope. Harvard University Press, MA
Farber D (2002) Sloan rules. Alfred Sloan and the triumph of General Motors. The University Chicago Press
Flink J (1988) The automobile age. MIT Press, Cambridge, MA
Kotler P, Armstrong G (1999) Principles of marketing. Prentice Hall
Rubenstein J (2001) Making and selling cars. The John Hopkins University Press
Sloan A (1964) My years with general motors. Doubleday, Garden City, NY (revised 1991)
Taylor A III (2010) Sixty to zero: an inside look at collapse of general motors-and the Detroit auto industry. Yale University Press

Chapter 4
The 1930s: Europe Behind in Marketing Strategies

Abstract In Europe, the history of the car industry and the evolution of marketing strategies took a very different course compared to the one taken in the U.S. In the first decades of the twenty-first century, Europe was influenced by Ford's model of mass production, yet maintained a tradition of craftsman-like automobile manufacturing which, to a certain extent, still persists to this day in the manufacture of luxury models and sports cars. Cars were mainly produced on a commission basis. They were an object of pleasure rather than of utility. There were few in circulation. Competitions triggered desires that only rich people were in a position to satisfy, and they chose the best manufacturers. During the 1930s, two trends emerged that left a lasting mark on Europe: a strong company identity and an indirect rivalry between products by the same manufacturer. In marketing, this was a period of design continuity by a single producer, and of a relentless search for beauty and diversity in terms of the visual. In those years, the main factors that influenced marketing were the emergence of new technologies, economic development (growing average income per capita), and the beginnings of mass consumption. Above all, though, it was influenced by the capacity of new entrepreneurs to understand the trends of potential demand and to draw upon technological and management knowledge tested and developed in other industries too.

Although in the modern conception the car was invented in Europe, from 1910 onwards various factors prevented companies on the old continent from rivalling the U.S. Two factors stood out: (1) Europe was the sum of many small parts, none of which, except for Great Britain, had a large enough car market potential to sustain mass production and initiate the virtuous cycle/spiral of high volumes, economies of scale, low costs and low prices, which had fed the growth in demand on the American market; (2) in the first decade of the century, the main European countries invested mainly in the production of weaponry and were on the verge of entering a disastrous war.

© Springer Nature Switzerland AG 2019
E. Candelo, *Marketing Innovations in the Automotive Industry*, International Series in Advanced Management Studies, https://doi.org/10.1007/978-3-030-15999-3_4

1   Object of Pleasure Rather than of Utility

With the U.S., it shared the characteristics of the first demand for automobiles and companies in the industry. In the early 1900s, the automobile was an object of pleasure rather than of utility. There were few cars in circulation. They were a product for rich people. They required servicing, often full-time, a driver, and even a mechanic. They could not travel too far from their trusted repair garage. Given that the first cars introduced onto the market often stalled, endurance races were the primary means of indicating the sturdiest in terms of construction and the most reliable.

In 1884, sponsored by *Le petit journal*, what is considered the first car race in history took place; covering 120 km from Paris to Rouen. It was presented as a "contest of horseless carriages". Although more than a hundred vehicles entered the race, only twenty-one of them were ready at the starting line. The race had two winners: Pahnard and Levassor and Peugeot.

Even car technology benefitted from the experience of races. Races triggered innovations not only in mechanics, but also in tyres, a very important component of the automobile due to the road conditions of the time. In *"Michelin Men. Driving an Empire"*, Lottman describes the initial scepticism and then interest of the French manufacturer, a leader in the production of pneumatic tyres for bicycles, concerning the races. Michelin did not participate in the Paris-Rouen race because he was doubtful about a rapid dissemination of the automobile and did not want to invest in an untested product, but following the great notoriety achieved by the event, he prepared and supplied tyres to several competitors in the second race, which took place in 1985 from Bordeaux to Paris, with a long return over 1000 km (Lottman 2003; Jemain 1984).

In the first decade of the 21st century, Europe was influenced by Ford's model of mass production, but it maintained a tradition of craftsman-like automobile manufacturing which, to a certain extent, persists to this day in the manufacture of luxury models and sports cars (Shimokawa 2012).

By 1915, in Great Britain, France, Germany, and Italy car production for civilian purposes had been discontinued. Everywhere, factories had started to produce arms, engines for airplanes, and military vehicles.

In the years immediately following the end of WWI, across Europe the types of car and production plants that had existed before the conflict were resumed. Production techniques for military materials had come on considerably during the war, and these advances were partly transferred to the production of automobiles. Various attempts were made to introduce modern techniques for the transition from artisanal production to volume production using assembly line methods, but without any notable success. The conditions for launching an expansion phase comparable to that which had driven mass marketing and production in the U.S. before WWI were in fact lacking.

In Europe, too, some of the most innovative companies had been established as bicycle and then motorcycle producers, a mode of transport that became increasingly popular after WWI. During WWI, for example, the German Wanderer supplied over half of the motorcycles used in the German Army. The bicycle remained the mode of transport for the masses. It offered access to the most desirable destinations that could not be reached by rail or other modes of public transport. It was also the most economical mode of transport. Most of the cars were used in the cities (Nieuwenhuis and Wells 2015).

Overall, the automobile spread from the richest classes to less wealthy ones, from high to low in the pyramid of incomes. Some companies that had supplied weapons to the armies found they had good profits at their disposal, which they used to convert their plants to the production of vehicles for civil use. Initially, due to a shortage of transport capacity, the demand was for light trucks.

In Europe, the recovery of the automotive industry began around the mid-1920s, but its evolution varied greatly from one country to the next.

2 Still Far from a Marketing Approach

Up until the 1920s, one cannot talk about marketing in the modern sense of the term. Cars were primarily produced on a commission basis. Races created desires, which only rich people could satisfy, and singled out the best manufacturers.

There was a considerable delay in the evolution of marketing strategies compared to the situation in the U.S., for at least two connected reasons. First and foremost, the middle classes did not have the means to purchase new cars and the few clients that there were wanted powerful cars built with modern standards. Secondly, no market in Europe, except for Great Britain, was sufficiently large to justify major investments in assembly plants.

Europe was divided into several national markets, separated by various types of barriers, including the wounds wreaked by the recently ended war, which had prostrated both the victors and the defeated. They were also separated by the peace treaty that had put an end to WWI, which had been highly punitive for the losers.

Car sales remained very low in each country. On the one hand, the limited sales volumes on the markets prevented companies from self-financing to cover new investments and, on the other, they restricted such companies from developing the economies of scale that could have led to a reduction in costs per unit and thus in prices, as well as to greater accessibility for consumers. Banks thus became involved in the management of companies (Nieuwenhuis and Wells 2015).

Despite its defeat in WWI, Germany was one of the first countries to vigorously resume automotive production. The 1920s was a period of product innovations: from the lubrication of engines to cooling systems. At the Berlin Motor Show in 1921, the first car series with a left-hand drive was presented. Auto Wanderer successfully participated in uphill and short-distance races. Races on local circuits were especially common in Italy, in cities. In 1922, the first Targa Florio took place in Sicily.

3   1920s–1930s: A New Beginning

During the 1920s, two trends emerged that left a lasting mark on Europe: strong company identity and indirect rivalry between products by the same manufacturer (Pellicelli 2014). To sum up, in marketing this was a period marked by continuity of design by a single producer, a relentless search for beauty and diversity in visual terms, and, once again, the use of racing as the primary promotional tool.

Strong firm identity. The demand came from rich clients who, in addition to performance in terms of speed and endurance, wanted cars that stood out clearly from the competition. Artisanal production favoured different interpretations of the external configuration of vehicles, motorisation, and suspensions. Each manufacturer, of those that had survived the crisis that struck the industry, had their own style. The differences were also influenced by the varying characteristics of the national territories and the fact that high tariff barriers offered protection against foreign competition.

A quick look at the photographs of some of the "national champions" of the time confirms this analysis. The Rolls-Royce Phantom (1925), Lancia Lambda (1922), Mercedes-Benz SSK 34 (1930), Alfa Romeo 1750 GS spider (1930) and Isotta Fraschini (1929) were the result of "product concepts" that were each noticeably different from one another (Chapman 2011).

These marked differences influenced the European industry's delay in developing mass production and mass marketing, and thus also the spread of the automobile. Since individual markets were closed to imports and as each market had its own leaders safeguarded not only by the State but also by strong firm identities, it was impossible for one or more European manufacturers to reach the volumes that might have triggered economies of scale and of scope comparable with those of the American industry (Clark and Fujimoto 1991; Ruppert 2011).

Indirect rivalry. One of the consequences of the strong identity of each company was less competition between manufacturers. Clients that preferred the product of a certain company because of its distinctive characteristics tended to remain faithful. If designers managed to propose other models, either following or anticipating the evolution of the industry, clients tended not to change manufacturer and to choose from the new proposals within the same product portfolio. This generated a sort of indirect rivalry between products by a single manufacturer. When the models of various companies (especially in the mass market) began to converge towards the same solutions, this was no longer the case.

These trends had certain consequences for marketing strategies: longer life cycles (because the competition was between models offered by the same company); continuity in the style of individual manufacturers (to maintain and strengthen the loyalty of existing customers); technological innovation and continuous improvements to performance, especially by manufacturers that chose speed races as their main form of promotion.

New competitors. In the second half of the 1920s, manufacturers in Great Britain and Germany had to confront the marketing strategies by Ford and GM. Initially, the two American manufacturers started to assemble parts and components imported from the U.S. in Europe, before then launching a phase of greater penetration by producing vehicles directly in Europe. They chose the two countries that promised the greatest development and with the largest industrial base: Great Britain (then the second market worldwide in terms of car production volumes, after the U.S.); and Germany (the European nation with the greatest industrial potential).

Ford built two large production plants, one in Great Britain (in Dagenham) and one in Germany (in Cologne), while GM acquired existing companies: Vauxhall in Great Britain, and Opel in Germany. The strategy adopted was to produce where they thought they could sell and, especially, produce as much as local demand seemed to call for. The principle had been theorised by Weber (1929) at the beginning of the century: "The best location for a factory is the place in which the sum between the costs of bringing in raw material, part and components from one side and the costs of transporting finished product to the final destination from the other side reaches the lowest possible point".

4 New Pathways Were Explored

The situation changed again in the 1930s, partly due to the economic crisis that had drastically reduced the demand for automobiles in Europe, but above all because the increasing hostilities between European countries were raising tariff and non-tariff barriers along national borders. Even in marketing and sales, new solutions were explored.

Germany remained the country that developed automotive production to the greatest extent. One of the main factors that triggered the diffusion of automobiles was without doubt the National Socialists' promotion of mass motorisation in their propaganda campaigns. The opening of the Berlin Motor Show in 1933 began with a speech by Hitler, in which he announced a programme of tax exemption for all new motor vehicles and the immediate construction of a network of motorways (*autobahn*). The launch of the People's Car project (Volkswagen), which was intended to offer the people a car suitable for transporting two adults and a child for the price of a motorcycle, was part of this policy. Nonetheless, before WWII broke out, only a few models came out of the factory in Wolfsburg. In 1936, the first crash test was organised in Germany.

New techniques in sales methods were introduced by companies that produced motorcycles and which later turned to the production of automobiles. In the space of a few years, DKW became a world leader in the production of motorcycles. It also played a pioneering role through the introduction of credit purchase sales. Efficient advertising strategies also made DKW popular beyond the German borders. In the late 1920s, DKW had one of the most extensive sales organisations. Its advertisement claimed: *"DKW—the Small Miracle: goes up hills like the others come down"*. An important source of financing for the expansion policy came from the State Bank of Saxony, which acquired 25% of the capital in 1929 In its advertising, DKW promised a pleasant drive, safety, low consumption, and comfort.

Everyone is delighted at the way DKW masters the road.

The DKW starts like lightning and gets you there in perfect safety.

A DKW fulfils the norm- low consumption, plenty of power.

Finish your journey as fresh as you started- space to spare in the DKW.

Soon or later you too will drive a DKW.

For a long time, Great Britain remained the second market in terms of volume after the U.S., but unlike the situation in the U.S. and other European manufacturing countries, market shares were divided between too many companies, which hampered the emergence of leaders with solid bases. In 1939, Great Britain operated with thirty-six manufacturers, six of which were focused on mass production, compared with three in the U.S. (the Big Three), France (Citroen, Renault, and Peugeot), and Germany (Mercedes-Benz, Auto Union and BMW), and one in Italy (Fiat). The proliferation of models and market fragmentation conferred great vivacity upon marketing but weakened the manufacturers that had survived.

French and Italian manufacturers shared a pioneering spirit that drove them to constantly seek out innovation in all areas, from products to advertising and distribution. Some also shared a passion for competitions. In France, Peugeot earned great popularity through its victories in the most difficult races, including Targa Florio and 24 h at Spa (Belgium). In 1934, it introduced the first convertible with an open-top roof onto the market (the 402 Eclipse Décapotable).

Unlike other French manufacturers (like Hispano Suiza), which started producing large, costly cars, Renault soon chose to focus on becoming the leading manufacturer of small cars. In the 1920s and 1930s, it produced a wide range of models. It was among the first European companies to export automobiles to the U.S.

While both Renault and Peugeot had been established at the end of the 19th century, having previously operated as bicycle manufacturers, André Citroen began producing spare parts for cars very late in the day. He earned considerable recognition when, in 1919, he introduced a car into the market at a considerably lower price than his competitors, but especially on account of being the first in France to offer a complete vehicle. Previously, the rule had been to purchase the chassis from one manufacturer and the body from another.

Citroen was a great innovator not only with products, but also with other marketing elements. He built a car specifically for women, which was very successful. He surprised Paris when an enormous Citroen sign appeared on the Eiffel Tower (it remained there for nine years). Unlike the other two French manufacturers, which were very careful with their profit and loss accounts, Citroen could not resist the temptation to spend beyond his resources, just to stand out. He went bankrupt in 1934.

In Italy, three manufacturers dominated the scene: Alfa Romeo, Lancia, and Fiat. Over the course of the 1920s, Alfa Romeo earned a reputation as a sports car manufacturer. It won the major races and the first world championship in the history of motoring, held in 1925. Product innovation was its primary objective. As such, each Alfa Romeo model had to represent a technical evolution compared to the previous one. Nonetheless, it was not so successful in terms of economic management, and ended up in the hands of creditor banks.

In the 1920s, Lancia's universally recognised masterpiece was the Lamda market, considered a synthesis of the genius of manufacturer Vincenzo Lancia. A photograph of this model with Greta Garbo at the wheel, a famous actress of the time, made it famous worldwide. Other highly successful models followed, such as the Dilambda and the Ardea. Lancia became well-known for continuously proposing innovative products and cutting-edge forms. Even the materials used were often completely new for automotive production. "The quality of the mechanics attracted the most well-to-do, demanding clientele" (Bruni et al. 2006).

In the years immediately after WWI, Fiat concentrated its automotive production on luxury and racing cars. The 519 was the flagship model of its fleet. It was the result of the latest technologies. At the same time, Fiat developed a series of racing vehicles that set important speed records. In the 1920s, Fiat underwent considerable expansion in Europe. It launched the Fiat Polski (1921) and a joint-venture for lorry production in Russia. The early 1930s witnessed the great success of the 508, baptised Balilla by the fascist regime and described as "an automobile finally directed at the people". The 508 was also a success in different European countries. To circumvent tariff barriers, it was produced in Germany by NSU, in Great Britain by Fiat England, in Czechoslovakia by Walther, and in Spain by Fiat Hispania. In France, it became the "6CV française".[1]

References

Bruni A, Clarke M, Paolini F, Sessa O (2006) L'automobile italiana. Giunti, Firenze
Chapman G (2011) Car. The definitive visual history of the automobile. DK Publishing, UK
Clark K, Fujimoto T (1991) Product development performance. Harvard Business School Press
Jemain A (1984) Les Peugeot. Lattès
Lottman HR (2003) Michelin men. Driving an empire. I.B. Tauris, London
Nieuwenhuis P, Wells P (2015) The global automotive industry. Wiley
Pellicelli G (2014) Le strategie competitive del settore auto. Utet, Torino
Ruppert J (2011) The German car industry. My part in its victory. Foresight Publication, Somerset West
Shimokawa K (2012) Japan and the global automotive industry. Cambridge Press
Weber A (1929) Theory of the location of industries. Chicago University Press

---

[1]"In fact, this is not really the case, given that the sales price is 10,800 l, enough to buy a house in the countryside; but it is certainly the lowest at the moment, and industrialists, retailers, employees, professionals, and especially women, to whom the advertising for the vehicle is explicitly directed, see the new Fiat as the much hoped-for automobile, finally accessible in terms of price, use, and maintenance" (Bruni et al. 2006).

Part II
The 1960s: Towards Convergence

Chapter 5
Back from the Brink: Europe's Surprisingly Strong Recovery

Abstract Throughout the 1940s, after the end of World War II (WWII), consumer expectations regarding automobiles remained the same as in the years immediately before the war, and the European car industry failed to propose new ideas. However, from the early 1950s, western European countries began to resume automotive production in a surprising way. In Germany, France, and Italy, companies capable of large-scale production emerged, such as Volkswagen in Germany, Peugeot and Citroen in France, and Fiat in Italy. They quickly established themselves as market leaders. The products or services that such leading companies developed often set the standard for other carmakers, thereby enabling them to achieve a prominent position among customers in quality terms. Their position as market leaders also provided them with some other advantages: better access to distribution channels; a strong and enduring image; and the means to achieve low production costs through greater economies of scale. In the mid-1950s, two differences in the marketing strategies of European carmakers began to surface: (1) a national identity that would gradually disappear after the Treaty of Rome (1960); and (2) strong differences between the strategies of mass market producers (such as Fiat, Volkswagen, and Opel) and those of high-end (e.g. Mercedes Benz) and niche specialists (such as Ferrari and Aston Martin).

Throughout the 1940s, after the end of World War II (WWII), consumer expectations regarding automobiles remained the same as in the years immediately preceding the war, and the car industry, for its part, failed to propose new ideas. Other problems arose, especially in Europe. Few people could afford to buy a car. Most European manufacturers did not have the resources to reconstruct the plants destroyed by the war and acquire new distribution structures. Many companies were forced to leave the market and close down for good, or they were taken over by the few others that had survived.

The recovery came about in the 1950s, a boom period for cars.

In Germany, France, and Italy, companies capable of taking advantage of the resumption of large-scale production emerged, such as VW in Germany, Peugeot and Citroen in France, and Fiat in Italy. They quickly established themselves as market leaders. The products or services that such leading companies developed often set the

© Springer Nature Switzerland AG 2019
E. Candelo, *Marketing Innovations in the Automotive Industry*, International Series in Advanced Management Studies, https://doi.org/10.1007/978-3-030-15999-3_5

standard for other carmakers, enabling them to achieve a prominent position among customers in quality terms. Their position as market leaders also provided them with certain other advantages: better access to distribution channels; a strong, enduring image; and a way of achieving low production costs through greater economies of scale (Pellicelli 2014).

Despite having come out of the war in a better position than its European rivals, Great Britain was weakened in the long term by governmental interventions that slowed down the consolidation of the car industry, leaving the plethora of manufacturers inherited from the 1930s virtually unchanged.

In the mid-1950s, two differences in the marketing strategies of European carmakers began to emerge:

- A national identity in marketing that gradually disappeared/after the Treaty of Rome (1960);
- Strong differences between the strategies of mass market producers on the one hand and those of high-end specialists and sport/luxury car producers on the other.

1   National Identity in Marketing

Years of closed borders along with governments' explicit desire to defend their national identities and different traditions helped maintain strong differences between the styles of European manufacturers for two decades. They chose different paths, but they were united by their search for ongoing mechanical improvements and their response to common external obstacles: high-energy costs; narrow, winding roads (motorways were still very scant); high population density in urban areas; and clients rendered demanding by their knowledge of the product.

In Germany, VW resumed productions with a focus on the mass market, Mercedes maintained the tradition of luxury production, and BMW followed it after going through a period of serious crisis. In this period, Audi was the manufacturer that many deemed the most innovative both in technological and marketing terms.

Audi's story is representative of the evolution and consolidation of the industry. During the economic crisis of the early 1930s, four brands—Audi, DKW, Horch, and Wanderer—together founded Auto Union. At the end of WWII, Mercedes Benz acquired the controlling majority in Auto Union, which it went on to sell to Volkswagen. Initially, Audi cars replicated VW products, but a turning point occurred when it introduced new product and marketing strategy concepts.

The VW group decided to create different positions in its market offer. With Audi (established through a temporary merger with NSU), it offered a clearly distinct brand positioned well above VW in terms of price, quality and image, with VM still focused on the mass market. Mainly, the VW group tasked Audi with achieving supremacy in terms of product and marketing innovation. Audi Quattro epitomised this new strategy: a product with a new style, based on new technology (permanent drive on four drive wheels), and with a "premium marketing" strategy based on its reputation as an innovative manufacturer and repeated victories in rallies and other endurance races, including the 24 h of Le Mans (Pellicelli 2014).

France and Italy (Fauri 1996) chose paths that differed in part from Germany's by concentrating on the production of small cars at prices affordable by the emerging middle classes (economy cars). The design engineers tended towards more spacious cars than those of previous decades. The Fiat 600 from 1955, for example, was a good-quality, small car that could nonetheless seat four people.

Renault had introduced the 4CV onto the market in the late 1940s. In the next decade, it successfully withstood the competition of the British Morris Minor and the Beetle by VW. The subsequent model was just as successful: the Dauphine. An attempt to enter the U.S. in the late 1970s, however, enjoyed little success. Dauphine proved unsuited to American traditions and was overshadowed by the emerging local production of compact cars.

In the 1960s, European mass market design engineers had more freedom in their choice of solutions. They focused more on compact cars intended for family transport, thereby opening up a new segment (family cars). The Italians were the first to introduce hatchbacks, aimed at achieving a higher load capacity. This solution then became the rule. In the following decade, the two most representative models were the VW Golf (which came out after the withdrawal of the Beetle), and the Fiat Ritmo.

During the 1960s, in Europe production grew rapidly. Germany and France led in terms of production volumes, followed by Italy and Great Britain. The former boasted between 2.5 and 3.5 million units produced, the latter two around 1.5 million. During that decade, the import barriers that protected national industries started to come down following international agreements leading Europe towards the Single Market. Consequently, competition between manufacturers became more intense. It became increasingly difficult for market leaders previously protected by tariff barriers to defend their market shares.

2 Strong Differences in Marketing Strategies

With the boom in the 1950s, a clear difference had already emerged between Japanese and American manufacturers on the one hand, and European ones on the other. While the former were almost exclusively mass market (or large-scale) manufacturers, the latter were either mass market manufacturers (Volkswagen, Opel, Fiat, and Peugeot, to name just a few), or "high-end" specialists. The latter could then be separated into premium manufacturers (Mercedes, BMW, Audi) and niche manufacturers (Rolls Royce, Bentley Jaguar, Porsche, Maserati, Ferrari, and many others). The various stages of consolidation of the automotive industry then placed manufacturers of all three types in the same product portfolio. This was the case for the Volkswagen group, which in subsequent years built up a portfolio consisting of both mass market brands (Volkswagen, Seat, Skoda), "premium" brands (Audi), and niche brands (Bentley, Lamborghini).

In the 1960s, substantial differences between the marketing strategies of three types of manufacturers began to emerge: mass market; "high-end"; and niche/sports cars/luxury cars.

(1) **Mass market** manufacturers necessarily had to choose mass marketing strategies based on: the expansion of market shares to create and sustain economies of scale and scope; maximum standardisation of parts and components; high levels of modularity; low and medium price ranges geared at targets with the greatest potential for overall demand; high production capacity to seize peak demand and reduce delivery times; advertising and purchasing incentives to avoid dropping below break-even levels in the recurrent stages of falling sales. Overall, such strategies were (and still are) vulnerable to the cyclical patterns of the economy and changes in consumer preferences, but especially to the "commoditisation" of products, which made it difficult to stand out from competitors and inevitably led to price competition and therefore profit volatility (Rubenstein 2014).

Innovations were (and still are) incremental. Among those that have changed production methods and influenced marketing strategies most significantly, one should recall the British Mini (the Austin-Morris group, then BMC), which revolutionised the market of small cars (considerable space, mainly thanks to the transversal engine and front drive). This was not the only example of innovation, however. Almost all manufacturers introduced models with a front drive and hatchback. The Fiat Uno, thanks to its attractive design by Giugiaro and driveability, sold more than 9 million units in just a few years.

(2) **"High-end"** (or "premium") strategies were based on a focus on differentiation, a prerequisite for charging a "premium price". The greater the differences perceived by consumers, the more the manufacturer could attract demand, preserve loyalty and charge a "premium price". Ways of producing at lower costs than competitors were generally limited, while there were numerous ways to differentiate. These ranged from product and service quality to brand image. For rivals, this made it more difficult to imitate (Nieuwenhuis and Wells 2015).

(3) **Niche** manufacturers had inherited the old artisanal production methods that had survived the rigorous selection process (Aston Martin, Ferrari, Lamborghini, Jaguar, Rolls Royce, and Bentley). They were able to impose their "product concepts" on the market and preserve their clients' loyalty intact. They controlled buyers' access by increasing prices and limiting production volumes to heighten the sense of exclusivity.

There was no shortage of drawbacks, however, as illustrated by the fact that almost all niche manufacturers lost their independence and definitively left the market. One disadvantage concerned relationships with suppliers. Purchasing small quantities of parts and components led to limited purchasing negotiation power. The greatest threat came from "high-end" manufacturers who, with their strategy of extending their range, were competing with niche manufacturers. Thanks to the advantage of distributing the investment costs over large volumes and thereby achieving economies of scale, they introduced their top-of-the-range models at competitive prices in market segments that were previously the exclusive territory of niche manufacturers.

References

Fauri F (1996) The role of Fiat in the development of the Italian car industry in the 1950's. Business history review 70:167–206

Nieuwenhuis P, Wells P (2015) The global automotive industry. Wiley

Pellicelli G (2014) Le strategie competitive del settore auto. Utet, Torino

Rubenstein J (2014) A profile of the automobile and motor vehicle industry. Business Expert Press, New York

Chapter 6
The "Golden Age" in the U.S.: From a Class-Based Market to a Personal One

Abstract In the 1950s, the cars by the Big Three, especially GM, placed style ahead of engineering, "form ahead of function", thus giving great importance to design. GM's designers were free to design cars as they saw fit, believed, and wanted. Advertisements frequently used words like "power", "mighty" and "bold", emphasising the sense of America's technological superiority at the time. Under the pressure of several European small-car specialists, such as Volkswagen, the market share of the Big Three declined in the late 1950s. In a few years, segmentation changed from being product-centred to being size-centred. The segmentation of the U.S. market by size helped the Big Three to defend their positions against imported cars, but hindered the path of the clear, class-based market positions of their full-sized cars. For as long as most households had only one car, the Big Three were able to maintain their traditional marketing strategies of positioning their products through prices to closely conform to the pyramid-shaped distribution of U.S. social classes. However, during the 1960s many American households started to own more than one car. When, in the 1970s, the energy crises drove Americans to buy smaller and more fuel-efficient family cars, Detroit's traditional, large family cars died out, and with them so did the strategy of creating a hierarchy of cars, differentiated by price, which appealed to people in every social class ("ladder of consumers").

Henry Ford had devised a means of selling American families their first car in the 1910s, and GM had worked out how to sell Americans a car to replace the first one from the 1920s onward. In 1930, 60% of American families owned a car. The percentage went down in the 1930s and 1940s as a result firstly of the Great Depression and then of WWII, but the objective of providing almost every family with a car resumed after the war (Rubenstein 2001).

After WWII, in the U.S. as well as Europe, for years demand for cars was based on the same expectations as in the pre-war period, and manufacturers needed neither to invest in innovation nor to change. For example, for around ten years after 1945, the Mercury Division of Ford Motor Co. still enjoyed moderate success selling the range of products it had designed almost ten years beforehand to the same, relatively homogeneous market, with one expensive/high-priced car, one medium-priced one and three low-priced ones. However, in 1956 it found itself in terrible economic

© Springer Nature Switzerland AG 2019   41
E. Candelo, *Marketing Innovations in the Automotive Industry*, International Series in Advanced Management Studies, https://doi.org/10.1007/978-3-030-15999-3_6

conditions. An article in *Business Week* on the terrible situation questioned: "Where are customers?" (Business Week 1956). There were multiple causes, but they were all rooted in the undisputed sense of superiority the Big Three had acquired in the preceding years, and which their contribution to the production of arms during a victorious war had only reinforced.

1   Importance of Design: "Form Ahead of Function"

The GM cars of the 1950s placed style ahead of engineering, "form ahead of function", "whimsy ahead of practicality" (Rubenstein 2001).

GM's designers were free to design cars as they saw fit, believed, and wanted. In the 1950s, they designed bodies that resembled jet planes and rockets to strengthen the public's perception of security in the technological dominance of their country (the first appearance of the iconic tail fin came with the 1948 Cadillac). Advertisements frequently used words like "power", "mighty", and "bold", emphasising the sense of America's technological superiority at that time.[1]

Many years later, critics were severe regarding "tail fins", but Americans at the time loved that type of car, particularly the style of GM, allowing the manufacturer to gain market shares by offering a design that clients considered attractive. On the contrary, Chrysler's post-war cars appeared tedious and old-fashioned (Gartman 1994).

"Despite critiques concerning the automobile and its design, place, and purpose in American society, the intense love affair with the car was unparalleled" (Heitman 2009). In their marketing strategies, carmakers did not think twice about riding this trend.

The "car for every purse and purpose" marketing strategy earned GM huge profits and a position of near-monopoly in the U.S. market. Had it decided to reduce its prices, it might have forced some of its rivals to leave the market. However, GM took the opposite direction and harnessed its position of dominance to raise prices. Its management knew that surpassing 50% of the market should have obliged the antitrust authorities to intervene. GM preferred to allow other carmakers to share half of the market (Rubenstein 2001).

With that kind of market dominance, the money poured in. Between 1950 and 1955, on average GM's return on investment stood at around a staggering 25%. GM was so large and powerful that the Justice Department wanted to break it up. Since its market share was increasing in 1956, in a down market, the government warned that it would need to take "extreme action" unless GM slowed down its

---

[1] According to Gartman, the "V-8 powered car of the era was nothing more than an opiate for hard-working Americans." He goes on to say that, "It served to lessen the rather harsh realities of a competitive capitalist system with its class structure, repetition, dehumanization, and repressive impulse [...] Therefore, during the 1950s, the car was a symbol and an expression of freedom at a time in American life when autonomy was in retreat" (Gartman 1994).

drive towards monopoly. Washington suggested that GM could better achieve these objectives by disinvesting one or two of its divisions. As well as dominating the car industry, GM produced 43% of the industrial vehicles in the U.S., built 60% of all diesel engines sold worldwide, and was the largest producer of refrigerators with the brand Frigidaire. The government's attention was well-deserved. "Besides being the biggest car company, GM was also the most proficient" (Taylor 2010).

Of the 88 American car companies that existed in 1921, only five were still present in the market by 1958. Besides the Big Three, these were American Motors and Studebaker. Of the five, GM, Ford, and Chrysler covered 90% of the sales. "Power in the auto industry had been concentrated, as if by natural selection and survival of the fittest, in the hands of three companies and one union. The model of corporate oligopoly and union monopoly seemed poised to stay in place for ever". "Meanwhile America was enjoying an historic post-war economic boom with Detroit leading the way" (Taylor 2010).

2 Segmentation by Size

Under the pressure of several European small car specialists, including Volkswagen for instance, the market share of the Big Three declined in the late 1950s. Arriving too late, the Big Three introduced smaller cars onto the market in the early 1960s. Each of them offered a compact car for the 1960 model year. Within a few years, segmentation had shifted from being product-centred to size-centred. GM's four main sizes were: subcompact; compact; intermediate; full-sized or "standard", most of which became larger during the 1960s. In addition to this, the Big Three also sold a variety of specialty models. The most successful were Ford's Mustang and GM's Chevrolet Camaro.

The segmentation of the U.S. market by size helped the Big Three defend their positions against imported cars, but it caused difficulties for the clear, class-based market positions of their full-sized cars (Rubenstein 2001).

During the 1960s, GM's old policy of "a car for every purse and purpose" was extended through a proliferation of models. Smaller cars appealed to young people with tight budgets, first-time buyers, urban commuters and two-car households, while larger ones appealed to the traditional American family. However, the policy of corporate twins, wherein virtually identical cars were sold under different brands, demolished the other axis of GM's long-term success: the clear positioning of its five divisions (Chevrolet, Buick, Pontiac, Oldsmobile, and Cadillac) in the market. With an exceedingly large variety of proposals, most of which were twinned, GM lost well-defined brand images for its range of products.

3 Traditional Marketing Strategies Died

For as long as most households had only one car, the Big Three were able to maintain their traditional marketing strategies of positioning their products through prices to conform closely to the pyramid-shaped distribution of U.S. social classes. However, during the 1960s many American households started to own more than one car. When, in the 1970s, Americans were driven by the energy crises into buying smaller and more fuel-efficient family cars, Detroit's traditional, large family cars died out, and with them the strategy of creating a hierarchy of cars, differentiated by price, which appealed to people of every social class ("ladder of consumers"). In the case of GM's strategy differentiated from luxury Cadillac at the top to the entry-level Chevrolet at the bottom.

References

Business Week (1956) Science can find their market. Business Week, 27 Oct, p 47

Gartman D (1994) Auto opium: a social history of American automobile design. Routledge, Chicago

Heitman J (2009) The Automobile and American life. McFarland & Co

Rubenstein J (2001) Making and selling cars. Johns Hopkins University Press

Taylor A III (2010) Sixty to zero. An inside look at collapse of general motors—and the Detroit auto industry. Yale University Press

Chapter 7
Lessons from the Japanese: The Third Paradigm

Abstract In the 1960s and 1970s, the marketing and production policies of Japanese manufacturers exerted a strong impact on marketing strategies as they saw market segments as being almost isolated from one another. "Product concepts" differed from one segment to the next. Together with the variability of purchasing behaviour, this led to highly varied marketing strategies and increased the intensity of the competition. Toyota pioneered a new approach that became known as the Toyota Production System ("TPS"), later called "lean production". The TPS combined the flexibility and accuracy of craftsmanship with the low cost of mass production. By using lean production methods, Japanese car manufacturer achieved production costs per unit well below those of European and American manufacturers, with greater volumes. They also succeeded in increasing the speed and efficiency of new product development, a significant ability in a competition in which time to market constituted an important advantage. The new Japanese production system had a major impact. It marked a radical new approach to the manufacturing process and created a paradigm shift in automotive manufacturing worldwide. In terms of marketing strategies, lean production made it possible to produce a variety of models on the same assembly line. By increasing the variety of the offering, it increased the capacity to open up new market segments.

In the 1920s–1930s, the Japanese market had been dominated by American manufacturers that imported individual parts (KDs) to be assembled locally. The situation changed, however, when the government, seeking to render the national industry independent, granted foreign companies sole local licensed production, and financed the establishment of Nissan, Toyota, and Isuzu (which mainly produced lorries).

The race towards the top positions in global production started shortly after the end of WWII. To develop its new ally, the U.S. decided on the transfer of technology to Japan, particularly during the Korean War (1950). Japanese car manufacturers quickly understood they would not be able to sustain the major capital investments needed to develop productions with American technologies. As such, they introduced a series of principles that were imitated by European and American manufacturers: the principle of the systematic, continuous reduction in costs; "*muda*" (the elimination of waste); *jidoka* (the injection of quality); and *Kanban* (the tag used as part

© Springer Nature Switzerland AG 2019

E. Candelo, *Marketing Innovations in the Automotive Industry*, International Series in Advanced Management Studies, https://doi.org/10.1007/978-3-030-15999-3_7

of a system of just-in-time stock control). In the early 1960s, through the Ministry of Industry ("MITI"), the government decided to support economic development by investing considerable resources in certain industries deemed strategic, including the automotive industry. This attempt was only partly successful, but economic growth was nonetheless achieved and generated greater demand for cars. At the end of the decade, Japan joined the group of countries with major car manufacturing companies. Its passage from a small manufacturer to a major one was inevitably accompanied by a change in the structure of the industry (consolidation and consequent lower number of firms), forms of competition, and marketing strategies (Nieuwenhuis and Wells 2015).

1 Door-to-Door Sales Replaced

Before WWII, in Japan carmakers exercised strong control over sales. Traditionally, most cars were sold on a door-to-door basis by sellers working for the manufacturer. High brand loyalty and limited turn-over of the sales force meant that customers were repeatedly contacted by the same company representative. This strengthened customer loyalty. Moreover, most vehicles were personalised based on the customer's requests, rather than being selected for sale from current stocks. After WWII, the door-to-door system was replaced with dealership network structures similar to those used in Europe. Dealers started having large stocks and offering pre- and after-sales assistance. Older people took a while to give up the old distribution model, while young people soon favoured visiting dealers before buying (Rubenstein 2014).

2 Strong Impact on Marketing Strategies

Clark and Fujimoto (1991) have illustrated how the specific nature of competition within the industry helped speed up and strengthen the capacity of Japanese manufacturers to compete internationally.

In the 1960s and 1970s, the Japanese market was characterised by a limited number of market segments with clearly distinct borders, within which strong competition arose. Manufacturers' marketing and production policies saw these segments as almost isolated from one another. "Product concepts" differed from one segment to the next. Together with the variability of purchasing behaviour, this generated considerable variety in marketing strategies and greater competition intensity. The number of companies operating in the market decreased, and the marketing strategies adopted by the strongest companies—Toyota, Nissan, and Madza—forced the weakest ones to leave the market.

In those years, the Japanese market was characterised by two types of marketing strategies, strong competition, and specific consumer behaviour (Clark and Fujimoto 1991).

- **Direct product rivalry**. In the Japanese market during the 1950s, manufacturers mainly looked at what was happening in Europe and the U.S. Certain manufacturers were inspired by American car models, others by European ones.
  For many years, market segments remained limited in numbers and clearly defined. Among themselves, the major firms competed product against product. Competition occurred within each of the main segments, without manufacturers having their own, individual line encompassing their various interventions. To a competitor introducing a new product, they responded by seeking to offer more, but always with the individual segment as their limit rather than the entire product portfolio. There was no company identity. "Product concepts" by the same company changed from one segment to the next and were therefore "unstable". There was no sense of Nissan-ness in products by Nissan.
- **Intense domestic competition**. In the 1950s and 1960s, competition in the internal market became more intense for two reasons: firstly, the rapid development of sales attracted many companies (in 1950, no fewer than eleven had mass production plants); and secondly, the aforementioned "product concept" instabilities had an impact, rendering every position a company achieved on the market constantly vulnerable. In other words, the policy currently adopted by major manufacturers when they present a large range of products did not exist, with each of them gaining strength by being part of the same product portfolio (shared platforms and parts, selective targeting, company image beyond individual product). This, too, helped intensify the competition and sieve out the remaining companies.
- **Instability of customer tastes**. Since the market was still in an initial stage of development, first-time car buyers had limited experience purchasing this product. They were greatly attracted by new products, with sales rising rapidly before then dropping off just as quickly under the pressure of new launches. Given that the strong competition continuously led to new products, short product cycles and frequent launches were a necessary move to survive on the market. In the 1960s and 1970s, the average length of product cycles in Japan was under five years, considerably lower than the levels in Europe and the U.S.

3 Lean Production: The Third Paradigm

Toyota interpreted the new situation established in the years immediately after WWII better than other manufacturers. Its approach became known as the Toyota Production System ("TPS"), later named "lean production" (Womack et al. 1990; Holweg 2007). TPS was developed by the mid-1940s and mid-1970s.

Lean production combines the flexibility and accuracy of craftsmanship with the low costs of mass production. Production was "driven" by demand down the line rather than being "pushed", as was the case with previous assembly-line systems, by the production rate coming from the foregoing team's work station in the line. With lean production, components are delivered just in time (with parts supplied only as and when they are needed) and every worker is encouraged to stop producing when it is discovered that something is wrong (Krafcik 1988).

Through lean production methods, Japanese car manufacturers achieved production costs per unit well below those of European and American manufacturers, which had greater volumes. They have also succeeded in increasing the speed and efficiency of new product developments, a significant ability in a competition where time to market constitutes an important advantage.

4 Greater Variety of Models

In terms of marketing strategies, lean production can produce a variety of models on the same assembly line. Before it was introduced, assembly lines could only handle one model at a time. Moving to a different car model, the closure of the line and expensive retooling operations were required.

The new Japanese production system, pioneered by Toyota, had a huge impact. Without entirely sacrificing economies of scale, it marked a radical new approach to the manufacturing process and created a paradigm shift in the world's automotive manufacturing. The TPS became the third production and marketing paradigm, different from both the ones introduced before (Henry Ford's assembly line plus Budd's all-steel body, and Sloan's multi-product, full-line segmentation).

The new technology did not remain secret for long, since Toyota, as per the characteristics of the approach adopted, disseminated the principles among its suppliers. Consequently, just in time and total quality management could be achieved only with the cooperation of all the suppliers. TPS was set to become widely-known and studied, first by Toyota's Japanese competitors, then by Americans and, lastly, Europeans. Increasing the diversity of the supply increased the capacity to open up new market segments, but it made it difficult to achieve large economies of scale, which had an effect on marketing (Shimokawa 2012).

In practice, the only market where the principles of large economies of scale could be achieved was North America, and national manufacturers reaped the benefits. When the borders for importing vehicles from Japan and Europe were opened, and after the successive introduction of transplants, it became increasingly difficult to achieve large economies of scale even in the U.S. Relatively small plants reduced the advantages of economies of scale but favoured a larger variety of products and new marketing strategies.

Sequential segment strategy On various occasions, Japanese carmakers displayed their capacity to come up with new marketing strategies in markets that were new to them. Often, new potential segments of demand emerged while the company's resources were limited, preventing them from taking full advantage of these segments. When this was the case, a sequential segment strategy could be a useful alternative, which Japanese manufacturers adopted on several occasions. The company entered the segment it considered to be the most promising. It invested the scarce resources at its disposal in this segment for a chance at success. If it succeeded, it directed the profits obtained from this success in the first segment to attacking the next leading segment in terms of capacity to attract. This sequence continues until the company has exhausted the opportunities offered by a given market. This strategy has been described by Best (1997) and Carr (2014), with a focus on the experience of Toyota.

Toyota. The Japanese carmaker entered the American market in the 1960s with the Corona at the low-price end of the market. After having successfully penetrated this segment, when the market became crowded with look-alike models from Honda, Madza, and Nissan and competition drove down profit margins, Toyota "sequentially moved to the next segment in terms of price and quality by adding the Corolla." In the 1980s, other models "were developed for higher price-quality segments" (Celica, Camry, Previa). In the early 1990s, Lexus was positioned at the high end, along with the Avalon and Supra. Over thirty years, in the U.S. Toyota created an efficient sequential strategy, and now it is present in almost all segments. "However when the business started in 1960, a multisegment strategy would have been impossible" (Best 1997; Carr 2014).

References

Best RJ (1997) Market based management. Strategies for growing customer value and profitability. Prentice Hall, NJ

Carr N (2014) The glass cage: automation and us. WW Norton & Company

Clark K, Fujimoto T (1991) Product development performance. Harvard Business School Press

Holweg M (2007) The genealogy of lean production. J Oper Manage 25:420–437

Krafcik JF (1988) Triumph of the lean production system. MIT Sloan Manage Rev 30:41–52

Nieuwenhuis P, Wells P (2015) The global automotive industry. Wiley

Rubenstein J (2014) A profile of the automobile and motor vehicle industry. Business Expert Press, New York

Shimokawa K (2012) Japan and the global automotive industry. Cambridge University Press

Womack J, Jones D, Roos D (1990) The machine that changed the world. Free Press, New York

Chapter 8
The Dawn of Globalisation

Abstract In the 1970s and 1980s, European, American, and Japanese carmakers shared two trends that signalled the first steps towards globalisation in the car industry: (1) product proliferation; and (2) convergence in marketing management methods. Consequently, a set of "universal" industry factors drove carmakers towards strategies of the same nature, wherein execution was often the ultimate success factor. The first step was to extend the product line (i.e. *product proliferation*) by introducing products that were not completely new but which marked an improvement on previous versions (i.e. *annual model*) into the market. By studying customers' new demands and expectations or based on technological improvements, companies were able to take the second step by introducing new products that completed their existing range, to increase their market share. Gradually, some factors started to generate convergence in marketing strategies: (1) strong capital intensity; (2) long product life cycle; (3) competition based on the proliferation of models; (4) every country having its "national champions" and the political sensitivity of the car industry; (5) the relative stability of base technology; and (6) the fact that the major carmakers assembled around 25–40% of the final product. The outer percentage was supplied by firms that sold to everybody. Differentiating strategies therefore became more and more difficult.

Two trends were shared by European, American, and Japanese carmakers, constituting the first steps towards globalisation in the car industry: product proliferation, and convergence in the distribution channels. Consequently, "universal" industry factors led to strategies of the same nature.

1 Product Proliferation: The First Waves

Towards the end of the 1950s, managed product proliferation became customary behaviour among major carmakers in America, Europe, and Japan. To cover the largest amount of the market possible, they began extending their product portfolios so as to occupy as many niches as possible. In 1957, in an article

© Springer Nature Switzerland AG 2019

E. Candelo, *Marketing Innovations in the Automotive Industry*, International Series in Advanced Management Studies, https://doi.org/10.1007/978-3-030-15999-3_8

in *Harvard Business Review*, Johnson and Jones (1957) argued that only with continuous product proliferation would firms be able to sustain long-term growth and that the more features marketers placed and sold in a product, the higher the value perceived by the customer and the more sales and profits they could generate (besides the advantage of economies of scale).

The car industry was among the first to follow the path of proliferation.

- The first step was to extend the product line by introducing onto the market products that were not completely new, but were an improvement on previous versions (i.e. *annual model*).
- By studying clients' new requirements and expectations or based on improvements in technologies, companies were able to take the second step by introducing new products that completed the existing range, so as to increase their market share. The objective was always to acquire customers' buying potential to the greatest extent possible.
- There was no shortage of proliferation strategies that involved simply adding features that were of little or no value for customers, but necessary to justify price increases.
- The fourth proliferation strategy involved using or adding new components to products with their own brand capable of creating distinction, again to justify higher prices. In substance, this anticipated what was to become the rule for "premium" brands from the 1970s onward: namely, the introduction of ABS, EPS, and many more features (Chandler 1963; Nieuwenhuis and Wells 1994; Shimokawa 2010).

2   New Rules of Distribution in Europe

The 1950s saw the start of the convergence towards similar forms of distribution in the main European markets. The introduction of regulations set by the European Community contributed to this, which became ever more stringent from the 1950s onwards. Such forms nonetheless differed greatly from one nation to the next, with concentration among dealers in some countries (e.g. Great Britain) and fragmentation in others (i.e. Italy).

Throughout Europe, direct distribution overseen by the manufacturer is very rare. The "one-tier" channel is much more common. The dealer acts as an intermediary between the manufacturer and the buyer.

For manufacturers, the presence of an intermediary in the distribution channel is advantageous inasmuch as they do not need to invest and, above all, it allows them to extend their presence on the territory, opening up greater access possibilities for potential customers. Disadvantages arise, however, when there is no direct contact between the producer and the consumer. The product might be top-quality, but there is no guarantee that the client will perceive it as such through the filter of the dealer (in the purchase and after the sale).

The 1950s saw the emergence of the distribution model that still exists in Western Europe to this day, with few variations. Two factors helped dictate the distribution structure ruled by manufacturers (Rapaille 2004):

(1) In the 1950s, dealer numbers increased rapidly and there was also a sharp rise in demand for automobiles. As always in the first stages of rapid expansion, the increase in dealer numbers did not assume the same quality in physical structures and in the organisation and supply of services. However, the dealer was central to the relationship between the manufacturer and the consumer, establishing contact between the two parties and assisting the customer both before and after their purchase.

(2) The quality of cars differed greatly from one manufacturer to the next and was not comparable to that of cars today. Post-purchase repairs were the rule rather than the exception.

Placing the main emphasis on the need to guarantee customers the safety of cars coming out of dealers' repair shops, and given the complexity of the product (which in fact had, and still has, few comparisons), manufacturers were granted the right by the national authorities (and, later, supranational ones) to assign dealers exclusive sales areas for their products and require that they use original spare parts and abide by standards for premises used for sales (design, layout, and size), staff qualifications, repair shop equipment, and the basic elements of services offered to customers. In practice, the same dealer could not serve more than one manufacturer, nor could cars be sold through other distribution channels. The few attempts to use alternative channels did not last long.

Given that demand surpassed supply in those years, dealers could achieve good profit levels. As such, they did not oppose regulations that clearly bestowed considerable power upon manufacturers.

3 Same Industry Structure, Same Strategies

The stronger drive towards convergence derived from the structure the car industry was assuming, which was common to all major markets and carmakers. Gradually, certain factors pushed toward convergence. These are the "universal" characteristics of the automotive industry (Nieuwenhuis and Wells 1994):

- Capital intensity. The automotive industry requires major investments. The main areas of investment centre on the development and production of internal combustion engines and the production and painting of body shells. High fixed costs command high production and sales volumes. Overcapacity is a major risk.
- Long product life cycles. Revenue comes years after the start of the first investment. The time extension of the cycle exposes the firm to large variations in demand and economic conditions.
- Competition based on the proliferation of models. To capture the demand of as many segments and niches as possible, major carmakers extend the range of their

models. This has three primary consequences: more intense competition; more complexity; and pressure on margins.

- Every country has its "national champions" and the automotive industry is politically sensitive. The rule of "too big to fail" and high exit costs frequently call for state intervention to actively support firms on the brink of bankruptcy. The result is unfair competition in certain markets and territories.
- In the automotive industry, most production does not involve radical innovation but rather continuous refinements. As such, base technology is relatively stable.
- Major carmakers assemble around 25–40% of the final product (they usually control the power train and marketing). The outer percentage is supplied by firms that sell to everybody. It is therefore difficult for carmakers to achieve distinction when the same parts, components, and even modules are sold to everyone.
- The design, production, and use of automobiles are subject to regulations concerning safety and environment impact that limit the strategic choices of carmakers.

References

Chandler A (1963) Chapters in the history of the American industrial enterprise. Beard Books
Johnson S, Jones C (1957) How to organize for new products. Harv Bus Rev 35:39–62
Nieuwenhuis P, Wells P (1994) The automotive industry and the environment. Woodhead Publishing, NY
Rapaille G (2004) Seven secrets of marketing in a multicultural world. Tuxedo Publications, NY
Shimokawa K (2010) Global automotive business history. Cambridge University Press

Chapter 9
The Secrets of Success

Abstract Few carmakers passed the selection process of the automotive industry's first decades of history. Many authors have dealt with the secrets of success. Tedlow (1996) identified a number of specific items that had a great impact on marketing strategies: the strategy of creating profit through volume; entrepreneurs having the talent and creativity to see new opportunities; the strategy of building an effective vertical system; the first-mover strategy of reaping high returns by building barriers against new entrants; the strategies of new entrants to attack those barriers; and success in the market being determined by how well the relentless change in competition and consumer behaviour was faced and managed. Some carmakers, led by exceptional entrepreneurs or managers skilled at coordinating mass production and mass marketing, obtained significant results in terms of profit and returns on investments. Others, attracted by these results, sought to enter new markets, but few succeeded in overcoming the barriers created by first-movers. The higher the barriers erected by first-movers, the longer they were able to hold onto their successes. In the car industry, one of the most relevant barriers is capital requirement.

Only a few companies came through the selection process of the first 70 years of history of the car industry. What were the secrets of their success? Many authors have tackled this question. According to Tedlow (1996), six factors had a profound impact on marketing strategies: (1) the strategy of creating profit through volume; (2) entrepreneurs having the creativity to see new opportunities; (3) the strategy of creating a vertical system through which "raw materials were sourced, production operations managed, and the product delivered to the end consumer"; (4) the first-mover strategy of reaping high returns by building barriers against new entrants; (5) the strategies of new entrants attacking those barriers; (6) success in the market being determined by how well the relentless change in competition and consumer behaviour was managed.

© Springer Nature Switzerland AG 2019
E. Candelo, *Marketing Innovations in the Automotive Industry*, International Series in Advanced Management Studies, https://doi.org/10.1007/978-3-030-15999-3_9

1   Profit Through Volume

What we now call the "skimming strategy" quickly became the predominant approach in the first decades of the previous century. The reason for this was simple. Companies can make a profit with the fewest possible investments and risks (as compared to those taken by first movers). Henry Ford's winning strategy was based on selling as much as possible, at the lowest possible prices, and making a profit through volumes. Entrepreneurs did the same in other industries as well: "They have not merely served the markets", "they have created them".

Sharing this logic, Sculley (1987) has compared Steve Jobs to Henry Ford. Through mass marketing of the automobile, Ford offered Americans at the time freedom of movement, while Jobs, over half a century later, gave them the greatest possible processing freedom through personal computers.

2   Entrepreneurial Vision

There has always been high demand in the transportation of people, including for mechanical devices such as the bicycle, the most widespread and economical means of transportation. However, there was no specific demand for a product like the Model T. "It was Ford, first and foremost, who had the vision of what a car should be." He did not follow the marketing strategy of others based on producing small volumes, selling at high prices, and seeking high profits. Entrepreneurs created mass marketing, and men of vision took the risks.

The ingeniousness of the first successful manufacturers lay in their capacity to transform high fixed costs into low unit costs, which could be obtained only if the fixed capital invested in plants and equipment was used as intensely as possible. "In other words, mass production demanded mass marketing".

If provided mass marketing the plants are used at maximum operative capacity, product costs per unit continue to go down. If unit costs go down, there can be a decrease in prices and an increase in the market share. If the market share increases, the competition structure changes and the industry moves from a monopolistic competition structure to an oligopolistic one. All the major carmakers today initially became oligopolistic in their original markets and then defended their positions when the industry moved towards globalisation.

3   Vertical System

How the vertical system was managed was another success factor. When it was managed correctly, cost savings offered competitive advantages. Once the pioneering phase of the automotive industry was over, to guarantee constant quality and uninterrupted supplies of raw materials and component parts, it was necessary to extend companies' boundaries towards the supply chain on the one hand, and towards distribution channels on the other, to ensure the relationship with the consumer. Managing the vertical system required ownership of the links in the chain at one or several levels. Ownership was equivalent to a secure defence, but there were various experiences of control without ownership, such as the franchise system in distribution.

In the following decades, almost all carmakers abandoned vertical integration and moved onto some combination of the administered, contractual, and corporate system.

4 First Movers and Entry Barriers

Some companies, led by exceptional entrepreneurs capable of coordinating mass production and mass marketing, obtained significant results in terms of profits and returns on investments. Other entrepreneurs, attracted by these results, sought to enter new markets, but only a limited number succeeded in overcoming the barriers created by first-movers. The higher the barriers erected by first-movers, the longer they managed to protect the successes they achieved. Porter (1985) defined barriers to entry as "the disadvantages that entrants face relative to incumbents".

The barriers varied in type, from economies of scale to brand identity, access to distribution, and absolute cost disadvantages. In the automotive industry, one of the most significant barriers is capital requirements. In the first two decades of the last century, to attack Ford a competitor would have needed to make a major investment in plants without having the accumulated learning effect and the brand identity of its rival.

5 The Competitor's Option

What strategies did carmakers adopt to successfully overcome the entry barriers? There are two options for contenders:

- *"Doing it better"*. They can adopt the same strategies as the incumbents. They have the advantage of being able to study and imitate strategies already underway and tested, seeking to implement them more efficiently (implementation option).
- *"Playing a new game"*. They can interpret market demand in a new way and devise new means of production or new distribution channels. If they are successful in redefining the game, new entrants might be able to push the incumbent into an unfavourable position in the market, rendering them unable to react.

In the early 1920s, due to the advantages achieved by Ford, GM could not have played the same game and produced something better than the Model T. Sloan wrote that, "No conceivable amount of capital short of the United States treasury" could have beaten Ford "at his own game" (Sloan 1963). As consumer behaviour had changed, GM decided to play a different game, segmenting the market and offering "a car for every purse and purpose".

6   Managing Change

Success belongs to those firms capable of managing change in line with the times. Even after the collapse of the Model T, Henry Ford did not understand that the market, consumers' needs and the basis of competition had all changed. He replaced the Model T with the new Model A, conceived as another "universal" car. He had the resources to compete with GM, but simply chose not to do so.

References

Porter M (1985) Competitive advantage. Creating and sustaining superior performance. Free Press, NY

Sculley J (1987) Odyssey. Pepsi to Apple. A journey of adventures, ideas and the future. Harper & Row, NY

Sloan A (1963) My years with General Motors, Dobleday, Garden City, NY. Revised 1991

Tedlow R (1996) New and improved. The story of mass marketing in America. Harvard Business School Press

Chapter 10
Marketing Science: The Beginnings

Abstract The initial stage of what today we call "marketing science" can be dated back to the 1950s, having first come to light in the U.S. It was founded by a group of individuals belonging to a university who were seeking to distinguish their research and publications from those of practitioners. The breaking point was the foundation of the Marketing Science Institute, the objective of which was to bridge the gap between academic studies on the one hand and the practical aspects of marketing on the other. Marketing science started to be considered as modelling marketing actions based on the disciplines of economics, statistics, operations research, and other related fields. Was there any impact on carmakers' marketing strategies? There is no clear evidence of a direct, fast impact on marketing strategies, especially in the car industry. However, the inception of Marketing Science indisputably paved the way for a conceptual marketing framework which, in that period, drove what is called the "marketing revolution". The major carmakers successfully assimilated many of the advances driven by marketing science. Such benefits included better-defined market targeting thanks to progress in motivational research and a deeper understanding of consumers' behaviour when faced with different pricing options, through a conjoint analysis. The main results were achieved through decisions based on econometric models or market experiments to measure advertising efficiency. These models could be classified into three streams of research: (1) models of advertising elasticity; (2) models of advertising carryover and dynamics; and (3) models of advertising frequency.

Convergence in the strategies of the major carmakers was also favoured by the spread of methods devised by marketing science, which became common heritage. The drive towards convergence derived from the heavy investments needed to compete, the long product life cycle of cars, and the prevailing uncertainty.

© Springer Nature Switzerland AG 2019
E. Candelo, *Marketing Innovations in the Automotive Industry*, International Series in Advanced Management Studies, https://doi.org/10.1007/978-3-030-15999-3_10

1   The Origins of Marketing Science

The initial stage of what today we call "marketing science" can be dated back to the 1950s, originating in the U.S. It was founded by a group of individuals belonging to a university seeking to distinguish their research and publications from those of practitioners (Vidale and Wolfe 1957). The earliest contributions "came from outside the marketing field, usually from faculty trained" in operations research/management science and working in engineering departments, rather than business schools (Winer and Neslin 2014). Some pioneering papers were published in the *Journal of Marketing* (Magee 1954), *Operations Research* (Vidale and Wolfe 1957) and *Management Science* (Anshen 1956).

A decisive change came about when a report from the Ford Foundation (1959), critical of business education in the U.S., called for a more rigorous approach to management and suggested that special attention be paid to research in the marketing field. The report *Higher Education for Business*, by Robert Gordon (Berkeley University) and James Howell (Stanford University), insisted that quantitative methods be incorporated into teaching in all management functions, from production to finance, and from R&D to marketing. The Ford Foundation reinforced its support by providing fellowships for business faculties interested in developing their skills in quantitative analysis and the social sciences (Winer and Neslin 2014).

Among those that took advantage of the programme were Robert Buzzel (Harvard University) and Philip Kotler (Northwestern University). Buzzell made a seminal contribution to marketing modelling research, which appeared in *Marketing models and marketing management* (1964). Kotler contributed later through *Marketing decision making: A model building approach* (1971).

Shortly after the Ford Foundation report, in 1962 the Marketing Science Institute ("MSI") was founded. Its objective was to bridge the gap between academic studies on the one hand and the practical aspects of marketing on the other. Marketing science started to be considered as modelling marketing actions using the disciplines of economics, statistics, operations research, and other related fields.

A number of retrospective articles about the origins of marketing science appeared in the Fall 2001 issue of *Marketing Science* by Steckel and Brody (2001).

What about the car industry? Was there any impact on carmakers' marketing strategies? There is no clear evidence of a direct, fast impact on marketing strategies, especially in the car industry, but the advent of Marketing Science doubtless paved the way for a conceptual marketing framework that, in that period, was entering what is called the "marketing revolution".

2 "Think Small": A Breakthrough in Advertising

During the 1950s, the Big Three considered competition by foreign cars as a minor injury. The reliability of those cars was extremely poor and spare parts took a long time to arrive by ship. Car imports made up a total of 1% of the U.S. market in 1955.

However, there was a growing interest in small cars as they appealed to the expanding number of U.S. households seeking economical, dependable second cars. The most successful of these was the Volkswagen 'Beetle', sales of which soared by more than six times during the 1960s (reaching an all time high of 1.3 million in 1971). Its American success was partly due to the revolutionary campaign by the advertising agency Doyle Dane Bernbach ("DBB").

Many consider this campaign to be the most inspired campaign ever for a motor vehicle, leading to a 'creative revolution' in the era of modern advertising through the use of techniques that challenged, surprised, and charmed the viewer (Vaknin 2008). DBB addressed the viewer with respect and intelligence. The first advertisement produced for VW was the "Think small" campaign in 1960, followed by "Nobody's perfect", which featured a Beetle with a flat tyre above the title. Another advertisement presented images of VW from 1949 to 1963, illustrating small changes and holding up to ridicule traditional carmakers' policies of planned obsolescence and annual model face-lifting.

In 1972, the Beetle overtook Ford's Model T as the world's bestselling car. However, as Henry Ford had done in the 1920s with the Model T, VW's management in the 1970s waited too long to replace the Beetle, reluctant to abandon a quarter-century success story.

3 Motivation Research on the Rise

The first motivation research and commercial use of opinion polling in the U.S. car industry occurred in the years just before WWII. Geared at predicting how human beings would behave, these initial attempts were rather limited in scope and expenditures for marketing research by businesses were somewhat small (Boyd et al. 1977). Nevertheless, there was some forward movement. In 1940, one of the first articles on motivation research was published in the *Harvard Business Review*. The author, Douglas McGregor, argued that, due to the intangibility of the motives, concrete empirical research was needed (McGregor 1940).

During the 1950s, greater interest among carmakers in exploring consumers' reactions to their proposals brought motivation research to the forefront. The dissemination of research was accompanied by a broad rationalisation of their qualitative basis and the contribution of experts in psychology. Ditcher, an Austrian psychologist, who had fled to the U.S. in 1938, distinguished himself by successfully applying the principles of psychology to resolving marketing problems. After becoming well-known for proposing new, winning solutions for selling Nestlé instant coffee and

Alka Seltzer, he also collaborated with some American carmakers (Boyd and Orville 1992).

By the 1980s, virtually all large firms were conducting formal in-house motivation research. Among these firms, major carmakers were at the very top in terms of expenditure. Long life cycles and major investments in the product development process were again the main drivers of this strong interest.

Given that motivation research is aimed at explaining how people establish their experiences, and how they learn and react to external stimuli ("marketing stimuli"), the growing complexity and speed of change in the economic and social field pushed researchers to exceed the limits of this discipline. In the 1960s, researchers started developing analytical models of what was then called consumer behaviour. It was a decade of great progress in marketing research.

In those years, the new conception of motivation research and its fusion with the concepts of market segmentation, targeting audiences, and positioning products, together with the computer revolution, also led to a profound, broad methodological development known as psychographics. Yankelovich (1964) was among the first to discuss the technique of measuring lifestyles and developing a classification of lifestyles.

Carmakers' marketers quickly grasped the value of information concerning preference awareness, reactions, emotions, prejudices, and other mental attitudes prevalent among the various demographic groups. Psychographic segmentation soon became the norm in the car industry.

Not one major carmaker was immune from the mapping fever.

4   Conjoint Analysis

In the early 1970s, the first conjoint studies were carried out, considered by some to be the most significant development in market research methodology (Green and Rao 1971; Green and Wind 1973; Johnson 1974). Once again, this was driven by the increasing complexity of marketing decisions. Many decisions are interdependent, for example those concerning positioning choices relative to competitors, pricing policy choices, and growth strategy choices.

Such decisions were largely made in conditions of uncertainty as to the development of the field, customers' responses, and competitors' reactions. For an efficient decision-making process, management firstly needed to understand, and then to be able to anticipate how customers would choose from several competing alternatives, and what impact each of these would have on the final result. In their decisions, consumers engage in a trade-off between several characteristics of a product or service. "Conjoint analysis is a set of techniques ideally suited to studying customers' choice processes and determining trade-offs" (Rao 2014).

In the automotive industry, too, conjoint analysis had various applications in marketing research for exploring various types of decisions, such as in target market selection and in the optimal design and pricing of new products. The main advantage of this method was the capacity to explore responses to "what if" questions directed at potential customers faced with hypothetical or real alternatives.

For example, in research concerning consumer choices regarding car prices, researchers collected the preferences for each hypothetical price solution described through the evaluation provided (by the consumer) of a series of attributes of the product and service to which the price referred. In general, they did not consider all the possible alternatives, but only a limited number of them. See the example about the car industry in the notes.[1]

As well as the vehicle sector, the method had various possible applications, from non-durable products, to industrial goods and financial services.

5 Advertising Effectiveness

Researchers quickly became interested in measuring the efficiency of modern advertising. Those spending money on advertising wanted to know how much to spend, on what media, with what contents, and for how long. Sophisticated, econometric and statistical methods quickly began to be used. Given that firms in the automotive industry are among those that spend the most on advertising (in relation to turnover, but also overall), it is logical that they were among the first to adopt the most advanced methods.

According to Tellis (2014), studies that have used econometric models or market experiments to measure advertising efficiency can be classified according to the following three streams of research: (1) models of advertising elasticity; (2) models of advertising carryover and dynamics; (3) models of advertising frequency.

(1) Models of advertising elasticity. Advertising elasticity is the percentage of variation in sales (or in market shares) per percentage point of variation in advertising spending. More exactly, it is defined as the elasticity of sales to advertising. Researchers evaluate advertising elasticity by examining the variations in sales or market shares owing to variations in advertising spending between one period and another, within a set timeframe.

A pioneering study on advertising elasticity was carried out by Palda (1964). Perhaps the biggest contribution came from Lambin (1976), who used data from different stages of the product life cycle, product categories, brands, and Western European countries.

---

[1] For example, researchers could ask a group of people representing the target in question to compare a car model offered at a price of $32,000 equipped with all-wheel drive, antilock brakes, and a sunroof with another model offered at $40,000 that also featured lateral airbags, a leather interior, and a high-quality audio system. Both models could be compared with a no-frills model priced at $22,000. The interviewees' responses could centre on choosing the combination of features and prices that the target considered to be the most appealing. "If the manufacturer has surveyed a representative sample, the conjoint analysis can provide fairly reliable data for determining the exact combination of features and price that the target market will find most appealing" (Nohria 1998).

Later studies by Assmuss et al. (1984) and Sethuraman et al. (2011) provided evidence that over a long period of time (20 years):

- Advertising elasticity had declined over time;
- Advertising elasticity was higher: (a) for durable goods than for non-durable ones; (b) in the early stage of the life cycle than the mature stage; (c) for yearly data than for quarterly data; (d) when advertising is measured in gross rating points ("GRP") rather than in monetary terms;
- Many of the results for short-term elasticity also apply to long-term elasticity, with a few exceptions.

These conclusions were all useful for carmakers, due to the size of their advertising budgets.

(2)   Models of advertising carryover and dynamics. Due to the nature of the product and the length of its life cycle, advertising carry-over attracted the attention of marketing management in the car industry.

The effects of advertising are not instantaneous. Rather, they are passed down and transferred over successive periods. Analysing advertising carry-over is important for at least four reasons (Tellis 2014). Firstly, the overall effects of advertising depend on the instant effects plus any carryover. If the carryover is significant, ignoring it leads to a drastic underestimation of the true effects of advertising. Secondly, if a "pulse" (an impact, or expense) of advertising has some carryover effect, it is preferable if the subsequent "pulse" is planned only after the effects of the first have run out. Thirdly, the length of the effects of advertising can have important implications for establishing whether the company should consider it as an expense or an investment, and if the treasury can allow it to be tax deductible. And lastly, the length of the effects of advertising can determine to what extent it is a barrier to other companies entering the market, or if it can create long-term habits (such as smoking).

Research into the effects of the frequency of advertising spending is equally important.

(3)   Models of advertising frequency. The expression "advertising frequency" refers to the number of times a consumer is exposed to an advertisement in a given period of time. Effective frequency refers to the optimum frequency that maximises the intended result of the advertiser, such as sales, profits, and price levels.

McDonald (1971) was among the first to publish studies on the effects of advertising frequency. Other contributions were put forward by Tellis (1988), who illustrated that: (1) "the effects of advertising were small and quite difficult to identify. In contrast, the effects of sales promotions were strong, immediate, and hard to miss"; and (2) that "brand loyalty moderated the effects of ad exposure. Buyers responded more strongly to brands to which they were more loyal". These conclusions were particularly useful for directing carmakers' investments in promotion. Another interesting conclusion to come out of other studies was that, "The effect of advertising frequency on consumers' choices is small relative to that of promotion". More predictable conclusions also emerged, however, such as that, "The optimum number of exposures of advertising varies widely by market, category, brand, and state of the consumer" (Tellis 2014).

References

Anshen M (1956) Management science in marketing: status and prospects. Manage Sci 2:222–231

Assmuss G, Farley J, Lehman D (1984) How advertising affects sales: meta-analysis of econometric results. J Mark Res 21:65–74

Boyd HW Jr, Orville CW Jr (1992) Marketing management: a strategic approach. Richard D. Irwin, Illinois (US)

Boyd HW, Westfall RL, Stasch SF (1977) Marketing research: text and cases. McGraw-Hill/Irwin

Buzzell R (1964) Marketing models and marketing management. Harvard University Press, Boston

Gordon R, Howell J (1959) Marketing decision making: a model building approach. Columbia University Press, New York

Green P, Rao V (1971) Conjoint measurement for quantifying judgemental data. J Mark Res 8:355–363

Green P, Wind Y (1973) Multiattribute decision in marketing. Dryden Press, Hinsdale

Johnson R (1974) Trade-off analysis of consumer values. J Mark Res 11:121–127

Kotler P (1971) Marketing decision making: a model building approach. Holt, Rinehart and Winston Series, New York

Lambin J (1976) Advertising, competition and market conduct in oligopoly over times. North-Holland Publishing Company

Magee JF (1954) Application of operations research to marketing and related management problems. J Mark 18:361–369

McDonald C (1971) What is the short term effect of advertising? Marketing science institute report no. 71-142. Marketing Science Institute, Cambridge, MA

McGregor D (1940) Motives as a tool of marketing research. Harvard Bus Rev (Autumn):42–43

Nohria N (1998) The portable MBA desk reference. Wiley

Palda C (1964) The measurement of cumulative advertising effects. Prentice Hall

Rao V (2014) Conjoint analysis. In: Winer R, Neslin S (eds) The history of marketing science. World Scientific

Sethuraman R, Tellis G, Briesch R (2011) How well does advertising work? Generalizations from a meta-analysis of brand advertising elasticity. J Mark Res 48(3):457–471

Steckel J, Brody E (2001) 2001: a marketing odyssey. Mark Sci 4:331–336

Tellis G (1988) Advertising exposure, loyalty and brand purchase: a two stage model of choice. J Mark Res 15(2):134–144

Tellis G (2014) Advertising effectiveness. In: Winer R, Neslin S (eds) The history of marketing science. World Scientific

Vaknin J (2008) Driving it home. 100 years of car advertising. Middlesex University Press, London

Vidale M, Wolfe H (1957) An operations-research study of sales response to advertising. Oper Res 5:370–381

Winer R, Neslin S (2014) The history of marketing science. World Scientific

Yankelovich D (1964) New criteria for market segmentation. Harvard Bus Rev 42(2):83–90

Chapter 11
Marketing Progress: A Never-Ending Story

Abstract During the 1960s, progress in marketing strategies driven by firms spurred on and led a strong evolution in marketing research, academic studies, and management literature. These included Sloan's *My Years with General Motors*, Levitt's *Marketing Myopia*, Ansoff's product/market matrix, Borden and McCarthy's marketing mix concept, Andrews' contribution to the SWOT analysis, the "learning curve", and the rise and fall of portfolio management, to mention just a few. All of them, directly or indirectly, had a considerable impact on the evolution of marketing strategies in the car industry, in which, due to the major investments and long product cycle, firms were desperately seeking tools and concepts to help them face uncertainty. The best contributions often came from managers who, during their professional lives, had introduced innovations and then successfully laid out their experience in articles or books. There was no shortage of academics and creative talent, such as Alfred Chandler, Theodore Levitt, and Robert Buzzell, who conceptualised management techniques or practices previously introduced and tested through the marketing function of carmakers. Consulting firms played a special role. Called upon to study situations and propose solutions, they achieved considerable success (more than academics) with the problems of car companies, in light of the actions taken by management and the results obtained by their proposals. Some of their best solutions became part of management's repertoire and were included in MBA textbooks, often using cases drawn from the car industry, such as the "BCG's growth-share matrix" and the GE/McKinsey matrix.

During the 1960s and 1970s, progress in marketing strategies driven by firms had spurred on and led a strong evolution in marketing research, academic studies, and management/business literature: these included Sloan's *My Years with General Motors*, Levitt's 'Marketing myopia', Ansoff's product/market matrix, Borden and McCarthy's marketing mix concept, Andrews's contribution to SWOT analysis, the 'learning curve', and the rise and fall of portfolio management, to mention just a few. All of them, directly or indirectly, had a great impact on the evolution of marketing strategies in the car industry, wherein, due to the major investments, companies were starved of tools for handling uncertainty.

© Springer Nature Switzerland AG 2019

E. Candelo, *Marketing Innovations in the Automotive Industry*, International Series in Advanced Management Studies, https://doi.org/10.1007/978-3-030-15999-3_11

Innovation in management emerges within the company, on the field. It is management that tests new solutions to old and new problems (i.e. the "car for every purse and purpose", Toyota Production System). When the solution has been tried and tested and demonstrated its practical value, the academy seizes it.

The dissemination of new solutions in management literature follows a cycle similar, in some ways, to that of a product. When a new management or analysis tool, a new idea (i.e. how to segment a market, or the ROI formulae family) goes beyond the embryonic testing stage and enters the development phase (being adopted by several companies), academics and those studying and writing about management start exploring it and situating it in existing literature and, sometimes successfully, seeking to give it a scientific position. They usually arrive after companies, but their work is useful for disseminating knowledge, facilitating learning for those that come after them, and paving the way for further progress. It is not always real innovation. As Sundbo (1998) observes, the history of marketing theories frequently "shows a process whereby old theories wither away, new ones arise and the old ones are revived in a new form".

The best contributions often come from managers who, in their professional lives, have innovated and then successfully laid out their experience in articles or books. Alfred Sloan (CEO of GM) was among the first to demonstrate this. Others followed, such as Ansoff (Lockeed's Vice President) and, in more recent times, Lafley (CEO of Procter & Gamble) (Lafley and Martin 2013). There is no shortage of academics that have conceptualised management techniques or practices (which is innovation, too) or demonstrated genius creative talent, such as Chandler (1963, 1990), Levitt, and Buzzell.

Consultancy firms played a special role. Called upon to study situations and propose solutions, they were very successful (more so than academics) in tackling companies' problems, what management was doing, and the results obtained as a result of their proposals. Some of their best solutions have been taken up by management and are included in MBA textbooks: "BCG's growth-share matrix", GE/McKinsey matrix.

1 Sloan's 'My Years with GM'

In the early 1920s, the "marketing revolution" was still a long way off. Written with the assistance of an editor from *Fortune* magazine, J. McDonald, and the young rising (at the time) historian, A. Sloan (1964), *My Years with General Motors* became a management classic. In this book, Sloan describes how and why he divided GM into self-governing, independent divisions (each with its own market and products) subject only to the financial and policy control of a small staff headed by the CEO. The information presented about market research illustrates how important this work was in reconstructing the role played by what was later defined as marketing in the organisation of a large company at the time.

In GM, Sloan wrote, market analyses were the responsibility of finance experts. They were entrusted, specifically, to "Distribution and Financial Staff". They had to provide the knowledge needed to evaluate the amount of resources to be assigned to each of the divisions and guide the pricing policy. Sloan recalls that, in 1923, following a slowdown in demand, the "Sales section" initiated research into the entire automotive market to evaluate the full potential for demand and, above all, to analyse "the probable effect of price reductions on the size of the market" (Sloan 1964).

2 Levitt's "Marketing Myopia"

In 1960, the *Harvard Business Review* published 'Marketing myopia', an article written by Theodore Levitt in which the author argued that, by defining their industry in terms of products rather than customer expectations, many companies were missing out on opportunities, and some eventually failed. There is a danger of "Marketing myopia" if a competitive business market is defined only in terms of product type rather than customer needs and expectations.

This approach had many implications for marketing strategies. Levitt gave evidence that:

(1) Rather than focusing their activities on a constricted definition of their products, firms should adopt a broad industry orientation. Thus, an automotive company should regard itself as being in the transportation business, in "mobility" in today's language, oil companies in the energy business rather than merely drilling and refining, film companies in entertainment, and so on.

(2) When management fails and its firm is overshadowed by competitors, "it usually means" that the product is not in tune with changes in customer buying behaviour, or with new and modified environmental conditions or innovation in complementary industries (Levitt 1960).

In American marketing literature, having lost sight during the 1960s of a more extended vision of its market, the railroad industry was frequently referred to as an example of "Marketing myopia". Railroad managers failed to recognise that their product was, simply, transportation. Consequently, while they were busy dealing with the railroad business (mainly based on the transportation by rail of goods, commodities, raw materials, and agricultural crops), automobiles and airplanes easily captured the passenger market, leaving the rails with only a very small fraction of the transportation market. By defining their industry in terms of products rather than customers' needs and expectations, the railroad companies lost a favourable condition for growth. The message may sound old hat now, but at the time it constituted a radical change.

Levitt's statement does not seem to have influenced the nascent automotive industry. Later, however, when a move towards diversification started to be considered, some carmakers were able to use the concept to their own advantage.

3 Ansoff's Product/Market Matrix

The evolution of strategic marketing was boosted by the popularity that strategic planning gained in the 1960s. In the mid-1960s, Ansoff, now considered as the father of modern strategic thinking, published a book that represents a landmark in management literature. In *Corporate strategy* (1965), Ansoff pointed out a crucial distinction between strategic planning and what he called strategic management. He also set out his reasoning on a series of rigorous processes designed to help managers reach strategic decisions, among which marketing management processes held a relevant position. Many of his ideas were later developed by famous writers, such as the concept of "competitive advantage" (by Michael Porter) and that of "core competencies" (by Gary Hamel and C. K. Prahalad).

In a further development, Ansoff disagreed with Levitt's position as discussed in "Marketing Myopia". He argued that a firm first needed to ask whether a new product had a "common thread" with its existing products. Ansoff suggested four categories for defining this "common thread" in the corporate strategy of a firm: (1) market penetration (the aim being to increase sales of existing products in current markets); (2) product development (organisations need to update their product portfolio to remain competitive); (3) market development (to find new markets for existing products); and (4) diversification (moving beyond existing areas of operation). These are the four combinations of current and new products within current and new markets known as Ansoff's product/market matrix.

In the 1990s, Mercedes Benz, the car division of Daimler, was a clear example of developing growth strategies based on the Ansoff matrix. *Market penetration*: the C-class (medium-sized family saloon) and the E-class (executive saloon) helped Mercedes Benz increase sales as well as reduce production costs, through greater production volumes. *Market development*: with the A-class small family saloon and the Smart car, Mercedes entered the small car market, while the relaunch of the Mayback repositioned the group in the super luxury price market. As well as achieving cost savings, the formation of Daimler Chrysler made it possible to develop lower-price brands worldwide through products such as the Chrysler Neon, and the Voyager. Nonetheless, problems arose concerning their quality, and the association with such down-market products tarnished Mercedes' reputation when it came to safety and quality. *Diversification*: the option of diversification was taken up by Mercedes, too. After an early entry into aerospace, its DASA defence subsidiary merged with the Franco-German EADS.

4 Borden's Marketing Mix Concept

In 1965, Neil Borden coined the expression "marketing mix" to describe the "tools and techniques" a company uses in a given set of circumstances to achieve marketing goals in its target market (Borden 1964, 1965). These tools and techniques were later summarised in four elements, known as the "four P's of marketing": Product, Price, Place, and Promotion. All these elements are deemed to be equally important, but in specific situations one or more of them may gain greater prominence. As for the automobile, in the first decades of the industry the product was unquestionably the most important element, but later it also maintained a special relevance among the four P's (McCarthy 1981, 2007).

Car marketing strategies of the time were not significantly influenced by the work of Coase, who published an article in 1937 on why companies exist, subsequently earning him a Nobel prize, Schumpeter (1942), who placed the entrepreneur at the centre of the growth process, or Penrose (1959), who published a book in 1959 in which she related a firm's growth to the resources under its control. As for Coase (1937), sixty years later, "several thinkers, seeking to understand how the Net is changing the firm turned to Coase's work" (Tapscott et al. 2000) (see forward in Part Four: Digital Age).

5 Andrews' SWOT Analysis

A conference held at Harvard in 1963 certainly had a much greater, longer-lasting effect, popularising the analysis model based on strengths/weaknesses (existing at the time) and opportunities/threats (likely, in the future) known as SWOT (Ghemawat et al. 1999).

It soon became a rule that after conducting a competitive audit and SWOT analysis, it should be easier to identify areas of key potential competitive advantage. Interpreting the results also offers guidance on how to revise a firm's marketing strategy. In other words, before formulating a marketing strategy, the market structure needs to be understood and described. It is necessary to carry out environmental scanning, that is, to understand how the competition, social trends, and government policies can hamper or support marketing campaigns, as well as to distinguish between those factors that can and cannot be controlled by management.

Developing these concepts, Andrews (1951) pointed out that in marketing strategies distinctive competencies and corporate resources, on the one hand, had to be matched with environmental conditions and trends and opportunities and threats/risks on the other.

The SWOT analysis was the subject of no shortage of criticism. The main one was that, ultimately, it inevitably relies on subjective estimates. Those in favour of SWOT responded that this did not matter: "The journey is more important than the destination" (Ghemawat et al. 1999).

6   The "Learning Curve"

The "learning curve" was first observed (according to literature originating in America) in the military aircraft industry and was first translated into management principles in the 1930s. It was demonstrated that increasing production volumes brought down the hourly cost of work needed to assemble each aeroplane. This observation was obviously not a new one.[1] Ford devised a strategy that called for total concentration on a single, "universal" car intended for everyone. Through the advantages of economies of scale and the learning curve, Ford reduced the price of the car and significantly widened the market in the process. What was new was the illustration of the practical consequences of the principle and the use of this principle in management and investment policies. The learning curve primarily affected the direct work content in a product or service.

The case of the automotive industry showed that using the experience curve as a key element to sustain a strategy choice presented many limitations.

In the 1970s, GM and Ford clearly had cumulative volumes in the U.S. well above those of Japanese companies in their domestic market. Why were the Japanese able to crack the U.S. market? Why were the Big Three forced to surrender a great deal of their market share to Japanese companies with smaller cumulative volumes?

As Buzzell and Gale (1987) observed, the competitive effects of cumulative volumes were misunderstood and overstated. Market share had very significant, but relative costs derived from other factors too. International comparative advantages have played a major role in the capital-intensive automotive industry. Compared to the U.S., Japan had lower capital and labour costs, which was especially relevant in terms of the inventory and logistics costs associated with the proliferation of models or options. The main reason why Japan was so successful in the automotive industry may be that "customer-perceived quality often outweighed price (cost) in their customer's purchase decision" (Buzzell and Gale 1987).[2]

---

[1] Each time production doubled, the amount of direct hours of work required went down by a regular percentage. In other words, the fourth airplane required only 80% of the hourly work required by the second, the eighth only took 80% compared to the fourth, and the fiftieth only took 80% compared to the twenty-fifth. The impact of learning on working hours and related costs could be predicted mathematically.

[2] In 1979, Michael Porter wrote an article for The Wall Street Journal pointing out the limitations of the experience curve as a device for formulating strategies (Porter 1979). By 1981, Walter Kiechel III (1981) had written about "The Decline of the Experience Curve" in a Fortune series that examined some of the major concepts being used to formulate strategies at the time. According to Kietchel, "... the curve is being consigned to a much-reduced place in the firmament of strategic concepts. With it is going a good bit of the importance originally attached to market share". Quoted by Buzzell and Gale (1987).

7 The Rise and Fall of Portfolio Management

In the 1960s and 1970s, the practice of calling upon the assistance of consulting firms spread among companies. One such firm, the Boston Consulting Group ("BCG"), took up the concept of the "learning curve" and redefined it as an "experience curve". Once again, they demonstrated that overall costs could be decreased by 20–30% every time production volumes doubled. This was due to economies of scale, organisational learning, and technological innovation.

Automotive manufacturers were among the first to turn to these new analysis techniques, driven by the amount of investment required to compete, the uncertainty continuously created by the turbulence of the environment and the reactions of the competition, as well as by the risks connected with long product life cycles (5–7 years).

The logic of experience curves suggested that it was not advantageous to try and finance each business through its development, or to finance the development of each business differently, independently of the overall business portfolio. Businesses with high growth or a high potential for growth were unable to generate sufficient cash to benefit from the market development, and were forced to relinquish shares of it. Mature businesses, on the other hand, with zero or very low growth, often generated more cash than they could invest with adequate returns. Many large companies understood (or rather, had the theoretical support to understand) the importance of a business portfolio and products comprising components with different competitive positions and different development potential.

- "BCG's growth-share matrix". What is known as "BCG's growth-share matrix" emerged. The two axes of the matrix are relative market share (or the ability to generate cash), and growth (or the need for cash). BCG developed the matrix to assist firms in deciding to distribute investments across "new" and "old" products. For a given portfolio of products, some take up more cash than they generate (cash users) and some generate more cash than they use. Marketing management needs to identify these to establish the right balance of products in the product mix.

Expressions such as "cash cows" (leaders in low-growth markets), "stars" (leaders in high-growth markets), "problem children or question marks?" (followers in high-growth markets) and "dogs" (followers in low-growth markets, static or in decline) became part of the common language of marketing.

Excess cash flows could be better used to gain or dominate market shares through investments in "stars" and "question marks?", for as long as the development of their markets remained high. If the two aforementioned businesses had been able to achieve and maintain a position of leadership, they would themselves have become "cash cows" when the market development slowed down and it was no longer necessary to invest (absorbing cash) to defend their positions. The "BCG's growth-share matrix" provided a strategy for managing the liquidity cycle.

This was the situation the major carmakers were embarking upon, especially in the 1960s, having adopted the proliferation of models as their marketing strategy (more models to occupy more segments or niches). The underlying idea was that gaining more market shares through economies of scale would lead to lower costs and greater returns. The premise behind this idea, and the consequent validity of the matrix, was clearly continuous market development. It therefore could not be applied to the very frequent situations of slowing or contracting growth due, for example, to more intense competition or an economic recession.

Other consulting firms took the same approach, proposing their own analysis methods and models to suggest how to chart and develop strategies. Each of the main consulting firms had its own matrix to offer.

- GE/McKinsey's matrix. In early 1970, following advice from McKinsey, General Electric divided its 140 profit centres into "strategic business units" ("SBUs"). It then asked McKinsey "to evaluate the strategic planning that was being written" about those SBUs. After analysing the problem, McKinsey established a nine-block matrix, with industry attractiveness on one axis and the competitive position, or business strength, of SBUs on the other. To evaluate the relative position of each SBU on the two axes, McKinsey suggested a set of criteria (including market share, investment intensity, product quality, and marketing expenditures).

Based on these criteria and using a defined weighted scheme, the SBU or products are placed on the matrix, which is divided into nine cells, three on each axis. The matrix became known as GE/McKinsey's matrix. It had a longer-lasting impact than other portfolio management proposals (McDonald 2013).

Shell's "Directional policy matrix" took a similar approach to GE/McKinsey's matrix (Shell 1975). In both matrices, the cells contain policy recommendations for businesses/products that fall within their boundaries, and the relative weighting of the factors in either axis is mainly based on subjective judgment. This was the source of great criticism.

All these matrices were used extensively in the automotive industry, but not for long. The popularity of such models decreased under the pressure of various (well-founded) criticisms. Consultants had devised a new product (named portfolio management) that did not avoid the fate of the life cycle of many products: introduction; development; maturity; and decline. The last stage was provoked by an unexpected, powerful event.

The oil crisis of 1970–73, with the consequent fall in demand, excessive production capacity and high inflation, highlighted the weakness of the concept of experience curves. Various authors, including Abernathy and Wayne (1974), illustrated that the reduction in costs due to experience curves was not automatic and that in many cases the systematic search for lower costs, like that based on experience curves, had reduced the capacity of firms to innovate and respond to competitors' strategies. Abernathy and Wayne also recalled that the obsession for continuous cost reductions and overlooking the change in consumer expectations had rendered Henry Ford vulnerable to the attack by GM.

Portfolio management in general was also widely criticised. This was for two reasons: (1) as it was based on an analysis of the current situation (the state of the economy, competition, and market trends), it was not suitable for business development choices, particularly in the automotive industry where the long product life cycle may go through turbulent times; (2) it placed an excessive emphasis on growth, while there were also opportunities in low-growth markets and even in declining markets, as repeatedly shown in the automotive industry; and (3) it required information that was difficult to obtain or time-consuming to collect.

The automotive industry soon recognised that the true role of models, such as BCG's matrix, was to help understand the real world. If used correctly, they can foster the first phase of a decision-making process. The goal of a model is to simplify the analysis of the external environment and/or the complexity of the firm's internal one. By definition, therefore, they all have some weaknesses. They can identify some of the critical factors and critical issues, but by no means can they safely generate an ultimate/optimal decision.

8 Towards a Big Shift in the Use of Data

Another progress that major carmakers fostered and were quick to take advantage of, with their research, was the new use of data to make marketing decisions. Advances in new computing technologies to gather large quantities of information and, above all, to quickly identify patterns of evolution allowed for a better understanding of the evolution in economic phenomena (consumer behaviour in particular) and gave marketing managers tools to make decisions not based solely on intuition. In an initial stage, the new tools were perfected and custom-built by single carmakers on course to devise the most refined solutions (such as Ford and GM in the U.S.). In a second stage, these solutions were commercialised by external specialist companies, and for this their use was widespread.

In an article that came out in *Harvard Business Review* with the title 'Analytics 3.0', Davenport identified these stages as the emergence of a new "business intelligence". Advances were made "in the form of operational efficiency" that improved performance, but they were still not understood as sources of competitive advantages. According to the author, it was not until the lead-up to the 2000s that one could speak of "competitive analytics" (Davenport 2013). We will return to the evolution of "Analytics" further in Chap. 16 (Table 1).

Table 1 From the mechanical age towards convergence

	Main drivers of change	Marketing strategies
United States		
1900–1920. Henry Ford's Model T dominates	Innovation in technology (moving assembly line). Economic growth	First time-buyers attracted by low-priced Model T. Mass marketing
1920–1950. GM rising as a market leader	Economic growth. Rising average incomes. "Ladder of consumers"	Segmentation; "car for every purse and every purpose"; branding
1950–1970. GM domination/near monopoly	Rising incomes, "a social class pyramid", evolution of consumer behaviour (the need to be different, search for emotions)	Style ahead of engineering, "form ahead of function". "Tail fins". Segmentation shifted from product-centred to size-centred. Traditional marketing strategy came to an end
Western Europe		
1900–1920. Object of pleasure rather than utility. One cannot talk about marketing in the modern sense until the 1920s	The arrogance of wealth, practicality (automobiles better than horses and bicycles), status. The market was primarily for the recreational and leisure purposes of the wealthy	To sell a small quantity of relatively expensive cars, high prices charged. To sell, producers must create a sensation, such as through high-speed races or long endurance trips across harsh terrain
1920s–1950s. Great differences in styles. Borders closed, different traditions, Europe fragmented into a collection of distinct national markets, isolated by governments through import quotas and high tariffs	Continuity of design in each firm. Racing as promotional tool and source of technological innovation. After WWII: devastation; lower living standards; basic transportation needs; production did not meet demand	Strong national identity. National champions. British-ness, German-ness, Italian-ness. Strong firm identity through continuity of design. Indirect rivalry among products by the same carmaker
1950s–1970s. Convergence towards the same marketing strategies	Economic growth; change in consumer behaviour; consolidation of the large number of low-volume manufacturers; surviving companies took advantage of mass production to produce large volumes of small cars. New paths were explored in marketing and sales as well	Convergence towards three types of strategies: • Mass marketing based on large market share to sustain economies of scale and scope, standardised parts and strong modularity. • High-end strategies based on differentiation to attain a "premium price". • Niche strategy by luxury and sports car producers

(continued)

Table 1 (continued)

	Main drivers of change	Marketing strategies
Japan		
Before WWII, the Japanese market was dominated by American producers who imported component parts (KD) to be assembled locally. The State's intervention changed the situation	Foreign firms were only allowed to produce locally under license agreements. The birth of the Japanese car industry was financed by the government: Nissan, Toyota, and Isuzu	Carmakers exercised strong control over sales, selling directly to consumers on a door-to-door basis. Selling without an intermediary strengthened customer loyalty
1960s. Few market segments with well-defined borders. Strong competition in each segment, product against product, tit-for-tat by major carmakers	Strong economic growth led to a growth in the demand for cars. Consequently, a great change in the structure of the industry (consolidation) and in the shape of competition took place	Strong impact on marketing strategies: • Direct product rivalry; • Intense domestic competition; • Instability of customer tastes
1970s. Lean production: the third paradigm in the history of the car industry	The major shift in marketing strategies was driven by lean production, which combined the flexibility and accuracy of craftsmanship with the low cost of mass production	In terms of marketing strategies, lean production made it possible to produce a variety of models on the same assembly line

References

Abernathy W, Wayne K (1974) Limits of the learning curve. Harvard Bus Rev 109–119

Andrews K (1951) The concept of corporate strategy. Irwin, Homewood

Ansoff I (1965) Corporate strategy. McGraw-Hill

Borden NH (1964) The concept of marketing mix. J Advert Res 4(1):7–12

Borden NH (1965) The concept of the marketing mix. In: Schwartz G (ed) Science in marketing. Wiley, New York, pp 386–397

Buzzell R, Gale B (1987) The PIMS principles. Profit impact of marketing strategy. The Free Press

Chandler A Jr (1963) Strategy and structure. MIT Press, Cambridge, MA (reprinted by BeardBooks)

Chandler A Jr (1990) Scale and scope. Harvard University Press, Cambridge, MA

Coase R (1937) The nature of the firm. Economica N.S. 4:386–405 (reprinted in Stigler G, Boulding K (eds) Reading in price theory. Homewood, Irwin, 1952)

Davenport TH (2013) Analytics 3.0. Harvard Bus Rev

Ghemawat P, Collins D, Pisano G, Rivkin J (1999) Strategy and the business landscape. Addison Wesley, Boston, MA

Kiechel III W (1981) The decline of experience curve. Fortune

Lafley A, Martin R (2013) Playing to win: how strategy really works. Harvard Business Review Press

Levitt T (1960) Marketing myopia. Harvard Bus Rev 45–56

McCarthy E (1981) Basic marketing: a managerial approach. Irwin

McCarthy T (2007) Auto mania. Cars, consumers and the environment. Yale University Press

McDonald D (2013) The firm. The story of McKinsey and its secret influence on American business. Simon & Schuster

Penrose E (1959) The theory of the growth of the firm. Basil Blackwell, Oxford

Porter M (1979) Experience curve. The Wall Street Journal, Oct 22 (quoted by Buzzell and Gale (1987))

Schumpeter J (1942) Capitalism, socialism and democracy. Harper, New York

Shell Chemical Company (1975) The directional policy matrix. A new aid to corporate planning. Shell International Chemical Co.

Sloan A (1964) My years with general motors. Doubleday, Garden City, NY (revised 1991)

Sundbo J (1998) The theory of innovation. Entrepreneurs, technology and strategy. Edward Elgar

Tapscott D, Ticoll D, Lowly A (2000) Digital capital. Harnessing the power of business web. Harvard Business School Press

Part III
The Electronic and Software Age:
A Rapidly Evolving Landscape

Chapter 12
The First Oil Shock: A Turning Point in Production and Marketing

Abstract The sharp increase in the price of oil—first in 1973 and then again in 1979—signalled a change in the structure of the competition, what to produce, and marketing strategies in the car industry. The oil crisis harmed the Big Three badly as American consumers turned to Japanese manufacturers. The oil crisis had pushed Americans to purchase cars on the basis of fuel consumption rather than segmentation based on social class. Having defined social class differences for half a century, the full-sized model market was now destroyed by Americans' preference for the smaller, more fuel-efficient Japanese cars. The mantra "listen to the voice of the customers" once again generated substantial results. The Japanese began to patiently build up a distribution network in the U.S. and, most importantly, to gain recognition for the quality of their products, their attentive after-sales service, the safety of their vehicles, and their reduction of harmful emissions into the atmosphere. In Western Europe, the consequences of the oil crisis were dramatic. Two trends came to light: (1) more cost-conscious consumers; and (2) the emergence of the German "premium brands". Audi, BMW, Mercedes, and Porsche developed a new "premium brand" strategy. During the 1970s and the early 1980s, the major carmakers began to extend the perimeters of their brand portfolios. They realised it was possible to serve more segments (through more brands) with great benefits in terms of lowering average costs, but to be able to do so, they needed to be able to sell at different prices in the various segments.

In the early 1970s, the global car industry was dominated by the Europeans and the Americans. During the previous decade, Japanese carmakers had considerably increased their production and saturated the internal market, but they had exported modest quotas and experienced various failures on the American market (including the Toyopet: see below).

The shock factor. The considerable rise in the price of oil—first in 1973 and then again in 1979—marked a change in the structure of competition, what was produced,

© Springer Nature Switzerland AG 2019

E. Candelo, *Marketing Innovations in the Automotive Industry*, International Series in Advanced Management Studies, https://doi.org/10.1007/978-3-030-15999-3_12

and marketing strategies in the automotive industry. The trend turned in favour of the Japanese industry.[1]

The oil crisis accelerated a change that was partly already underway. Not only did the way of competing change, but so did society, the markets, technologies, and consumer purchasing behaviour. Not everyone understood what was happening. Paradigms that had persisted for decades—Ford's Model T, Sloan's car "for every purse and every purpose", and lean production—influencing the selection of information and decision-making criteria, made it difficult to recognise that a way of producing was waning. Sunken costs (accumulated with the enormous investments in plants and equipment of the previous decades) had reduced flexibility. Many thought, or at least hoped, that the past would return, but it was not so.

The 1970s was also a period of rapid inflation in the U.S. and Western Europe. As prices rose in some years at a "double digit" rate for the first time in more than a generation, managers paid increasing attention to the impact of inflation on traditional methods of fixing prices and measuring and reporting profitability. It suddenly became evident that depreciation charges, based on original acquisition costs, were insufficient to cover the rising replacement costs of plants and some other fixed costs. Reporting profits on an "inflation-adjusted" rule became standard practice. The influences on marketing were profound.

The logic of accounting principles recommended including costs calculated on depreciation quotas reassessed on the basis of the replacement cost of the plants in the price of vehicles sold. However, selling at rising prices driven by inflation threatened to reduce demand. For this reason, too, few manufacturers kept up the profitability levels of the previous years.

1 The Big Three Lost Their Market Share

The U.S. reacted to the rapid escalation in the price of energy by reducing its dependency on petroleum imported from the states of the Persian Gulf and increasing its petroleum "strategic reserves" (stored in caverns along the Louisiana and Texas coast) by more than fourfold. The policy that had the greatest impact on the automotive industry was the government's decision to require carmakers to achieve a more efficient use of energy. All carmakers selling more than 10,000 vehicles a year in the U.S. had to meet corporate average fuel efficiency (CAFE) standards. As a result, the average fuel efficiency increased during the late 1970s and early 1980s.

Many other decisions regarding production functions improved fuel efficiency, such as decreasing the weight of car components, aerodynamic drag and operating inefficiencies. In addition, the extended use of radial tyres reduced fuel consumption,

---

[1]"... in 1973, in protest at American and Western support for Israel, a number of Arab oil producers began to boycott the supply of oil to the West, principally to the U.S. Within just two months the price of crude oil quadrupled" (Holweg and Oliver 2016).

as did the design of sleeker front ends. Aluminium replaced steel in body panels, hoods, and plastic bumpers.

Caught by surprise. The Big Three were badly affected by the oil crisis. Ford posted heavy losses in the late 1970s, Chrysler almost went bankrupt, and GM did not understand that the market was changing in a dramatic, lasting manner. GM's management thought that minor adjustments would be sufficient. Having refused to restructure in the late 1970s and early 1980s, GM entered the last decade of the century with outdated products.

The marketing strategies of American carmakers confused consumers as for many years they sold a mix of recently downsized products and older models. By way of reaction, American consumers turned to Japanese manufacturers. Toyota, Nissan, and Honda all steadily increased their sales by offering smaller, gas-efficient vehicles, meanwhile the Big Three lost their market share.

In the U.S., import quotas for smaller Japanese vehicles with lower consumptions increased, reaching 10% of the market in the space of just a few years. The oil crises had driven Americans to purchase cars based on fuel consumption rather than their social class. The full-sized model market, which for half a century had defined social class differences, vanished as Americans preferred the smaller, more fuel-efficient cars made by Japanese manufacturers (Rubenstein 2014).

2 "Listen to the Voice of the Customers"

In Japan, the oil crisis of the early 1970s arrived while the manufacturers' strategy was changing once again. As the internal market was almost saturated by that point, with companies having invested heavily in plants and relied upon economies of scale as one of the pillars of their capacity to compete, to keep up their development rhythms and profitability levels they were forced to turn to exports. Even with the government's support, Honda, Nissan and Toyota made a greater effort to penetrate foreign markets. Together, they exceeded a million vehicles exported in a year by several times. The majority went to the U.S. This success did not only derive from the improved fuel efficiency of the vehicles. Adopting a clever "blind-side" attack strategy, for some time already the Japanese had been focusing on the market segments overlooked by the American industry: small cars.

Brick by brick. The mantra "Listen to the voice of the customers", which over the past two decades had become an integral part of the Japanese tradition, driven by strong internal competition, once again produced significant results. Looking further ahead, Japanese manufacturers began to patiently build up distribution networks in the U.S. and, above all, to earn recognition for the quality of their products, their thorough after-sales service, the active and passive safety of their vehicles, and their reduction in harmful emissions into the atmosphere. Their "customer first" policy, with a focus also on environmental problems, had been overlooked by American carmakers up to that point. The Japanese avoided repeating the errors that Toyota had made with the Toyopet model in the second half of the 1950s, for instance. Toyota had in fact not chosen an optimal time to enter the American market, since in those years Americans were attracted by large, powerful cars (the oil crisis was still a long way off), and the margins for the small number of dealers that agreed to sell Toyota cars were still modest. Over the next twenty years, Toyota succeeded in patiently establishing a carefully selected network of dealers, present only in the main urban centres (so as not to waste resources) and with models that benefitted from the image it had acquired in the meantime, as a reliable, high-quality manufacturer.

Leap-frogging strategy. The American experience also involved a successful leap-frogging strategy (in an attack mode: overtake each other in turns). If you are performing poorly in all segments of the market, "try to design a leapfrog move that will not only negate your competitive disadvantage, but also create a sustainable competitive advantage and maintain differentiation in your market" (Buzzell 1964). This was the marketing strategy that the Japanese introduced with the compact-car market in the U.S. Japanese carmakers penetrated the North American market thanks to their fuel economy and low-cost superior position. However, in the early 1970s, the Japanese had a reputation for making poor-quality cars. To adjust their position in an upward direction, they leap-frogged Detroit on several key attributes other than fuel economy and low cost and shifted into better value positions. Toyota's advertising tag lines for the 1970s were *"You asked for it – You got it – Toyota"* (you got a car that met your basic transportation needs), and in the 1980s, *"Who could ask for anything more?"* (Buzzell 1964).

3 Western Europe: Two Lasting Effects on Competition

In Western Europe, the consequences of the oil crisis were dramatic. Governments intervened by limiting energy consumption in various forms. Gasoline was rationed. Some countries took drastic measures. Many of them banned all but emergency motor vehicle travel on Sundays. New regulations required carmakers to improve upon fuel efficiency. The production of motor vehicles, as well as of steel and other energy-dependent industries, plunged rapidly. Overcapacity, which was already a persistent issue, increased dramatically. Thousands of jobs became at risk. In Western Europe, the effects of Japanese competition were less of an issue. Imports from Japan did not surpass 5% of the market. However, the fall in demand due to increased energy costs placed more than one manufacturer in a state of crisis (including British Leyland in the UK, for example), accelerating consolidation in the industry.

Two trends destined to leave a profound mark emerged in the purchasing behaviour of consumers and in competition between manufacturers. The consequences for marketing strategies were considerable.

More cost-conscious consumers. Because of the oil crisis, Western Europe entered a period of austerity during which, for most of the population, frugal living and consuming (avoiding any excess, being sparing in the use of certain goods) became a prevalent behaviour. The middle classes saw their income and purchasing power decrease and later stagnate. Even wealthier states such as France and Germany were badly hit.

As a result, consumers became increasingly concerned about receiving more value for their money. In the process, the car industry was deeply affected. More and more people were interested in buying low-cost cars besides wanting greater fuel-efficiency. Similarly, governments were forced to introduce a new wave of austerity. An ageing population, spiralling health-care costs and the pension burden started to impose extensive budget cuts.

Germany's "premium brands" emerged. The 1973 oil crisis clearly rendered the sports cars and luxury cars that had dominated their respective niches in the previous decades obsolete. They were lacking in terms of safety, fuel efficiency, and new technology. Threatened by the supremacy of Japanese manufacturers, experts foresaw that only a small number of companies worldwide would survive. Instead of pursuing the strategies of the Japanese, Audi, BMW, Mercedes, and Porsche developed a new "premium brand" strategy. By setting new benchmarks, they brought about a revolution in the car industry. They managed to position their "premium brands" in the high-end segments of the car industry through new technologies, a strong image, and pricing. In *Premium Power* (2006), Rosengarten and Stuermer, retracing the history of the German "premium" brands, demonstrated how each one established differentiation by interpreting three concepts in a different way: innovation, design, and speed.

4   Trade Wars: The Rise of the Japanese "Premium Brand"

In the late 1970s, the crisis in Detroit drove American manufacturers to ask the government for protection from Japanese imports. Similar tensions arose in Europe, where French manufacturers led the protests. The effects were nonetheless limited because the Japanese responded by entering the higher price segments and building plants in markets into which they had previously exported.

In particular, the decision by the U.S. to impose "voluntary" limits based on the number of vehicles rather than their value drove the Japanese to move to the high-end segments of the market. They did not manage to overtake the "premium" German manufacturers, but instead of reducing, the pressure on American producers grew. In their exports to the U.S., the Japanese in fact focused on vehicles offering higher profits and introduced "premium" brands: Toyota with Lexus, Nissan with Infiniti, and Honda with Acura.

In Europe, Japanese manufacturers responded to the limits set by import quotas by building production plants in France, Great Britain, Spain, and Eastern countries. Having started from scratch in 1960, by the mid-1980s, so in less than thirty years, the Japanese had conquered a quarter of the world car market for passengers. In the same period, the world share of American carmakers dropped from 50 to 25%.

5   The World Car's Lack of Success

Over the course of the 1970s, more than one manufacturer toyed with the idea of the "world car", a vehicle based on a shared platform and components sold in several markets worldwide, with just a few adjustments. Examples included GM with the "J platform" project, Fiat with Palio, and Ford with Focus. They were not as successful as anticipated. Each continent and market had its own requirements and its own level of evolution in terms of demand.

What were the reasons? In the larger, more homogeneous markets—Western Europe, the U.S., and Japan—convergence towards the same marketing strategies was increasingly evident, driven by the limits of differentiation. At the same time, however, the different regulations concerning competition, along with differences in climate, culture, traditions and geography rendered many markets unique and made the possibility of offering the same products partly futile. The development of micro cars was supported in Japan by the policy of taxing car ownership based on engine size. In Europe, the high price of fuel supported the dissemination of diesel engines, while in Brazil low ethanol prices boosted the development of flexible engines capable of burning either petrol or ethanol. The three-wheeled car was an almost exclusively Asian phenomenon.

6   New Models of Segmentation

During the 1970s and 1980s, along with the downsizing and restructuring of carmakers in response to the escalation of oil prices, changes in consumer behaviour created new challenges in decisions about how to segment the market and position a brand or product on the market in relation to competitors. With the increasingly intense competition, the capacity to differentiate products, which had become relatively stable and efficient in the previous two decades, was rapidly and continuously weakened. At this stage, the input of consulting firms was considerable, particularly for the automotive industry.

With its own clients in the sector, the Boston Consulting Group (BCG) started to explore market segmentation in the early 1970s. The problems encountered and solutions proposed are illustrated in *The Boston Consulting Group on Strategy* (Deimler and Stern 2006) in which Tilles (1974),[2] a partner of BCG, gave a useful description

---

[2]Tilles (1974) described one of the criteria of segmentation indicated by BCG as follows: "For differentiated products, the basis of segmentation is the combination of the features built into the product and their cost/price ratio". "For example, Cadillac, Torinos, and Volkswagen are all very

of what was happening: "A differentiated product remains a differentiated product only until the emergence of the first follower. After that, it begins to behave as a commodity". As time goes by, all products tend to lose their capacity to stand out from the competition. The same rules applied for automobiles.

As a result, there were two paths open to companies: either the high-end/niche market, or the mass market. With the rapid evolution of the market in the early 1970s, companies in the vanguard were thus faced with a choice between: (1) limiting production volumes, differentiating products, charging high prices to support the high costs of being specialist producers; or (2) becoming high-volume producers, with low costs, standardised products, and low levels of differentiation.

There were no obvious responses to what was best. Much depended on the strategic orientation and availability of financial resources of the individual company. German "premium brand" producers—Audi, BMW, Mercedes Benz, and Porsche—chose the first option, as we said earlier.

Marketing strategies developed further when the major carmakers began to extend the perimeters of their brand portfolios. They understood that it was possible to serve several segments (through several brands) with great benefits in terms of lowering average costs. To do so, however, they needed to be able to sell at different prices in the various segments. "Cost to the customer must match value in each segment. Different value requires different prices to cover different costs" (Tilles 1974). Volkswagen was the main protagonist of this strategy. After acquiring Auto Union (later renamed Audi), it repeatedly expanded its brand portfolio.

Towards the beginning of the 1980s, advances in the information economy gave a new impetus to segmentation techniques. Information technology made it possible to identify, trace and analyse product and service transactions. Together with the progress in production flexibility, companies in the automotive industry were able to use information about consumer behaviour to personalise both their communications and their products and services. As such, they could serve segments as small as one individual consumer.

Marketing thus continued to evolve with the advances in segmentation. The complexity of managing the marketing mix increased due to the range of responses to the increasingly intense competition. In particular, pricing policies became more difficult, while the objective was to ensure the consistency of such policies. The car industry continued to be divided up into several segments, since consumers traded up and down in relation to price. What had long remained a relatively stable mass middle market was overturned.

"Segment-of-one" marketing. In the previous decades, marketers had succeeded in narrowing their focus and producing automobiles designed for specific segments. Over the course of the late 1970s, thanks to the new information technologies, it was possible to carve out a segment tailored to the individual. The segment-of-one combined two previously distinct concepts in one working relationship: information

---

different in their price-feature relationships and for that reason do not compete directly with each other. The segmentation of markets for differentiated products rests on the relationship between the cost features to the producer and the value of features to the customer".

in retrieval, and service delivery. On the one hand there was a proprietary database of consumer preferences and their purchasing behaviour, while on the other was an approach that uses the information base "to tailor a service package to individual customers" (Winger and Edelman 1989).

The central nucleus of "Segment-of-One" marketing is the capacity to track, know, and understand individual consumer behaviour. Thanks to advances in electronics, the opportunities to capture data had increased and storage costs and management costs for large databases had both gone down drastically.

From the perspective of competition, Winger and Edelman note that the consequences were considerable. While economies of scale in production were subject to continuous erosion in many industries, "Segment-of-One" marketing re-established scale economies in information management, in the service package, and in distribution. The consequence, the two authors conclude, was that the competitive advantage moved towards companies that simultaneously "own the market and are able to satisfy individual customers' needs".

The power of targeting. Dawson (2003), making specific reference to the situation in the U.S. regarding the evolution of targeting, offers a different interpretation, which is certainly more analytical than Tiller's.

By the beginning of the 1970s, economic stagnation had taken hold in the U.S. once again. In the context of "stagflation" and increasing competition between companies, market segmentation methods became less suited to the objective of identifying new consumer expectations. A new quantum leap in targeting methods was required. It soon appeared.

Dawson recalls that, before the 1970s, progress had already been made in market targeting, starting from the observation that people tend to be grouped into clusters with the same attitudes and behaviour. Marketers selected two or more demographical or economic characteristics and plotted them on a Cartesian graph indicating the point of intersection between two traits, characteristics, or dimensions, for example the age and income of the potential buyer. Using a mathematical or "eyeball" method, they identify an area around these points. In this way, researchers could establish a broader representation of the potential target as they could segment on the basis of more than one dimension, characteristic, and several lifestyle or demographic attributes.

Dawson also notes that a researcher, Jonathan Robbin, came up with the idea of using the targeting data held by companies—"survey results, customer response information and so on"—and mathematically combining/"crossing" such data "with publicly generated information available from both the Census Bureau and the U.S. Postal Service", two large sources of data that marketers tended not to use. Robbin called this new targeting method "geodemographics".

Thanks to these advances, companies were able to reach not only restricted groups of people, but also individuals ("hypertargeting"). A new generation of faster, more powerful computers enabled markets to "zero-in on ever smaller niches of the population, ultimately aiming for the smallest consumer segment of all: the individual."

This proved that it was both possible and effective to develop targeting by drawing upon previously unutilised reserves of demographic information about the public on the one hand and the capacity of computers to store masses of information and calculate complex scientific correlation that escaped the ordinary human mind on the other. In substance, new bases were established for targeting.

This increased companies' capacity to understand and thus to change or take advantage of the environmental, behavioural, demographic and financial conditions that influence consumer choices. Ultimately, Dawson concludes bitterly, it increased the possibility of marketing convincing consumers to accept innovations or further product proliferation.[3]

> **Self-inflicted wounds.** In the early 1990s, GM had only just come through a market crisis when it decided to give a new order to segmentation once again. Instead of considering cars as a continuum from small- to large-sized and low to high-priced models, GM decided to design specific models for specific client segments, based on age, income, and lifestyle requirements. It called this technique a "needs segmentation analysis", which experts immediately took as the revival of a policy from previous years. Each brand needed to be tailored to appeal to precise groups of customers.
>
> Unfortunately, this idea erased all design continuity between brands because each model was designed in a vacuum. It also marked another failing in market research. One result was in fact the production of cars such as the Aztek, which "seemed to have dropped to earth from outer space". The myriad features of this vehicle had been carefully market-tested, but "the whole was less than the sum of its parts" (Taylor III 2010).

7 The Risks of a Proliferation of Product Lines

Many examples demonstrate that the car industry was at the cutting edge in segmentation techniques. In 1973, in *Marketing Strategies*, a collection of managers' experiences in the industry published by the Conference Board (a business research organisation), Gail Smith, General Director of Advertising and Merchandising for GM, wrote that "Marketing planners at General Motors do not consider the segmentation approach the best or the most fruitful approach way to conceptualize the automobile market." Many of the "basic assumptions" of the segmentation theory were considered to be too restrictive. In particular, the attempt to identify clusters of

---

[3]In *The Consumer Trap. Big Business Marketing in American Life*, Dawson illustrates how "big business marketing campaigns penetrated and altered the lives of ordinary Americans."

Despite the aforementioned progress in market targeting, the management of major carmakers often continued to follow old traces and commit grave errors. GM is one example of this.

potential clients by their demographics, current needs and wants or market behaviour patterns is refuted because they were not sufficiently stable to be "*a means of planning our product alignment or marketing strategies*". Smith explained that at GM, the automobile market was in fact considered to be "a constantly shifting interplay between several rapidly changing forces". Consequently, at GM, market segmentation implied the design, production and distribution of additional product lines. In essence, research into market segmentation "is to discover whether the product line can be expanded". GM's marketing strategies were therefore based on certain products geared at stable, or relatively stable, segments, and on the search for the new, the emerging, the potential, or the latent.

The other side of the coin revealed major problems in terms of cost efficiency, however. When it came to advertising, GM had to face some "of the most serious shortcomings of the segmentation approach". Smith recalled that the larger the medium, the lower the cost per thousands, and the more specialised the medium the higher the cost per thousands.

As the objective of segmentation is to reach homogeneous clusters of people and as people's needs and wants vary constantly, the result is that the use of a specialised approach implies relatively high-cost media.

Through the continuous expansion of product lines, companies in the automotive market sought to occupy the largest possible number of segments (to distribute fixed costs across the largest possible base). The consequences were not only an excessive proliferation of models in the industry, but also a general decrease in advertising efficiency in the budgets of carmakers and the increasing impact of the advertising share on prices.

General Motors. In the early 1970s, for example, General Motors produced 24 makes of cars, each one different from the next. "Each of these 24 makes had important variations; we produced four-door sedans, convertibles, station wagons, and two-door hardtops. Further, we have 'series within makes' based on options and trims that can create significant and noticeable differences between models and some make." From this standpoint, General Motors produced 139 cars which are discernibly different to the consumer and therefore, can be expected to appeal to a wide variety of consumers. This presents a most difficult problem for the advertising manager. "We cannot simply devise 139 separate advertising campaigns, each with its own media schedule, and each with its special messages and product appeal. Such campaigns would have to be small in scope and would be lost in a sea of other advertising vying for the public's attention. The largest of advertising budgets cannot cover a product array of that scope". Source: Gail Smith in *Marketing strategies*, Conference Board 1973.

8 Organisational Buying Behaviour at the Forefront

In the 1960s and 1970s, models of organisational buying behaviour were the subject of considerable interest. In the car industry, too, selling to organisations (firms, rental companies, public organisations) was becoming increasingly important. The profits achieved through such sales were lower than those that came from sales via dealers, but this was compensated by the advantage of being able to establish medium- to long-term agreements with clients/organisations for large production volumes and thereby render volume planning more stable.

Bridging the gap. In those years, marketing literature extensively discussed the main differences between the buying behaviour of individuals and that of organisations. Firstly, the demand of organisations is derivative. They buy what they understand their clients want or foresee what they might request. Secondly, the purchasing process of an organisation is rarely the responsibility of one single person. Usually, it is a group decision. And thirdly, given the financial dimension of the purchase, which is generally high, and the complexity of the decision, which involves financial, legal and technical aspects, the decision-making process is usually long.

The 1970s saw the publication of many articles on organisational buying. Robinson et al. (1967), in one of the most original contributions, divided the buying process into straight rebuy, a modified rebuy, and new purchases. "The more novel the purchase, the greater the degree of uncertainty and the more people are involved in the purchase decision". The buying centre expression was coined.

Two models became "noticed": the Sheth model (1973), and the Webster-Wind Model (1972). Both were later considered to be applicable to a "wide spectrum of organizational contexts as they identified the most important variables in the process" (Webster 1984).

The complexities of the organisational models mentioned above limited the attention paid to marketing practice, but their contribution to a more rigorous approach was undisputable. Together with the advances in marketing literature, the buying behaviour of organisations in the car industry became increasingly significant and complex.

The emerging role of fleets. In the main markets, sales to organisations started to make up a significant portion of sales of new cars in response to changes in client demand, similarly to what was happening in other distribution channels. In US, compared to other markets, the evolution was fast and structured. It gradually assumed forms that were later adopted in Western Europe and, to a lesser extent, in Japan.

- The sale or rental of cars to employees—manufacturer internal fleet sales—was the first form to develop, especially in markets where the manufacturer had a high number of employees.
- Initially, the phenomenon of sales to networks of dealers was limited to the U.S., since in Europe and Japan the organisation of dealers into networks was practically nonexistent.

- Rental fleets made up the largest share of sales to organisations, especially on account of the development of global brands Hertz and Avis. Hertz (founded in 1918) gradually consolidated its position as a market leader. It was at the centre of frequent phases in ownership. In 1967, the Hertz Corporation became a wholly owned subsidiary of RCA. Avis was founded in 1948: it became famous in marketing strategy for having adopted (in 1962) the corporate motto "We Try Harder". This signified: "we know we are the second largest car rental company in the United States (after Hertz), but this is why you should choose us, because we are fully committed to overtaking our rival".[4] Rental fleets also contributed to the introduction of new information technologies in the car industry. In 1972, Avis introduced Wizard, the first computer-based information and reservations system to be used in the U.S. car rental business.

From the 1960s to 1970s, practices concerning relations with manufacturers were introduced that still exist to this day. Manufacturers often used surplus stocks for rental fleets. From the outset, manufacturers had adopted financial mechanisms geared at involving networks of dealers in managing excess vehicles in the pipeline, which could be likened to current "zero kilometre" used cars. To support the sale of new vehicles, manufacturers often undertook to buy back used cars from rental fleets once a certain amount of time had passed (once they had been replaced by new vehicles in the rental fleets offered to clients). Since the residual value of cars tended to go down on average with the increased market shares acquired by fleets, some manufacturers intervened to directly manage second-hand vehicles. The relationship with rental fleets became integral to manufacturers' development strategies. Some, when they had excess liquidity and were wondering how to invest it, chose to purchase direct or indirect shareholdings in rental companies (i.e. Ford-Hertz).

References

Buzzell R (1964) Marketing models and marketing management. Harvard University Press, Boston
Dawson M (2003) The consumer trap. Big business marketing in American life. University of Illinois Press
Holweg M, Oliver N (2016) Crisis, resilience and survival. Cambridge University Press
Robinson PJ, Faris CW, Wind Y (1967) Industrial buying and creative marketing. Allyn and Bacon
Rosengarten P, Stuermer C (2006) Premium power. The secret of success of Mercedes Benz, BMW, Porsche and Audi. Palgrave
Rubenstein J (2014) A profile of the automobile and motor vehicle industry. Business Expert Press, New York
Sheth J (1973) A model of industrial buying behaviour. J Mark 37(4):50–56
Smith G (1973) Market segmentation and advertising strategy in marketing strategies. In: Bailey E (ed) Conference board in marketing strategies. A symposium. 1974. The Conference Board, New York

---

[4]The slogan was used for 50 years before a re-branding in 2012, during which Avis acquired a new motto: "It's Your Space".

Taylor A III (2010) Sixty to zero. An inside look at collapse of General Motors and the Detroit auto industry. Yale University Press

Tilles S (1974) Segmentation and strategy. In: Deimler MS, Stern CW (2006) The Boston Consulting Group on strategy. John Wiley & Sons, Hoboken, New Jersey, pp.139–140

Webster F (1984) Industrial marketing strategy. Wiley, New York

Webster F, Wind Y (1972) A general model for understanding organizational buying behaviour. J Mark 36(2):12–19

Winger R, Edelman D (1989) Segment-of-one marketing. The Boston Consulting Group, Boston, MA

Chapter 13
Mass Customization: Another Marketing Breakthrough

Abstract In the 1990s, driven by further changes in technologies and customer expectations, marketing strategies in the car industry achieved another breakthrough. In their pursuit of excellence, some carmakers successfully pioneered the shift towards mass customization. In the early 1990s, two stages of the value chain were partly mixed/blended together: mass production, and mass distribution. This change was driven by three key factors, namely: (1) limits to the mass production process; (2) divisions into social classes, age differences, differences in lifestyle, and inequalities in the distribution of income, which reduced the homogeneity of the markets and drove towards dishomogeneity; and (3) instability of demand. There was a transition from a sellers' market to a buyers' one, and from markets "ruled" by sellers to markets "ruled" by buyers. In place of homogeneous markets, standardised products and long product life cycles, a new market structure emerged, consisting of heterogeneous and fragmented markets, a variety of products and shortening product life cycles. The internet significantly increased the possibilities of mass customization. Carmakers were able to put together components and modules as requested by customers at the last minute before delivery. While the main goal of mass production was to develop, produce, and market automobiles at prices low enough "that nearly everyone could afford them", the main goal of mass customization was to develop, produce, and market affordable automobiles and services so that "nearly everyone would find exactly what they want".

1 Three Main Factors of Change

In the early 1990s, two opposing management methods were combined/blended: mass production, and mass distribution. The change was driven by three primary factors: (1) limits to the mass production process; (2) reduced market homogeneity; and (3) unstable demand (Pine and Gilmore 1997; Gardner and Piller 2009). These factors had originated in the previous decades, but gradually began to have a greater impact.

(1) **Limits to the mass production process**. Maintaining efficiency in production processes is a prerequisite of mass production. This requires stable inputs,

© Springer Nature Switzerland AG 2019   95
E. Candelo, *Marketing Innovations in the Automotive Industry*, International Series
in Advanced Management Studies, https://doi.org/10.1007/978-3-030-15999-3_13

production processes and outputs. In the 1980s, this stability was obtained by management controlling these variables. This was no longer the case.

- *Input*. Labour costs and the price of raw materials and components were the main factors. The relative stability of labour costs and their occasional reduction were achieved through constant improvements in the productivity of such labour. Stable sales prices were also obtained through a high level of vertical integration (See Chap. 3). When productivity started to decline in the 1970s, the capacity of mass production to lower costs was weakened.[1]
- *More flexibility needed*. The standardisation of products and manufacturing processes, broken down into small specified tasks, conferred stability upon production processes, which also contributed to the high specialisation of employees and equipment. Market fragmentation made production processes more flexible and consequently reduced the capacity to keep production volumes stable.
- *Output*. Management's control of the effects of variations in demand was achieved, in the short term, by manoeuvring stocks. To increase efficiency, production rates needed to remain as stable as possible. These rates were set based on demand forecasting. If demand fell below certain levels, in the short term production rates were kept stable by manoeuvring stocks. If the drop persisted in the long term, or reduced to a greater extent, adjustments needed to be made to the production capacity, including through ill-fated dismissals. Considerable oscillations in demand regularly occurred on the market (particularly during the first and second oil crises: in 1973 and 1980, respectively).

(2) **Reduced market homogeneity**. Homogeneous markets are an important factor in the initiation of economies of scale. The oil crisis of the early 1970s marked the start of a period of change in customers' expectations. Social class divisions, differences in age and lifestyles and income distribution inequalities reduced market homogeneity and drove towards dishomogeneity. While customers' needs and wants became uncertain, it was difficult to find a single product or a limited number of products offered by the same company. It was more likely for a number of different products to be offered by many companies seeking as many profitable niches as possible.

(3) **Instability of demand**. To create the prerequisites of economies of scale, demand needs to be structurally stable (that is, in terms of the elements it is made up of), with no sharp fluctuations, strong upswings or downswings. As

---

[1] The significant increase in oil prices in the early 1970s introduced a high level of instability, which had a negative effect on costs. For example, lowering costs through economies of scale made it possible to lower prices and thus offered a greater possibility of extending marketing shares. The main variable was the cost of labour per unit. In order for this cost to go down, either real salaries needed to decrease or labour productivity needed to increase. From 1970 onward, this no longer occurred as it had in the past. In particular, productivity dropped on all fronts or increased at a slower pace than in previous decades. One possibility for reducing prices and extending market shares, which was essential for fostering economies of scale, was thus removed (Pine II 1993).

long as the demand for a product is stable and predictable, production levels can also be stable and predictable. On the contrary, when demand breaks into many fragments, it becomes difficult to predict volumes, production planning is troubled, and economies of scale degenerate.

Two factors in particular helped reduce the stability of demand and forced companies to adopt new marketing strategies: an increase in buyers' negotiating power, and new technologies providing a more varied offer.

- Markets already saturated or headed towards saturation, or even in crisis due to the recurrent recessions, shifted the negotiating power between the parties involved in the transactions. There was a transition from a sellers' market to a buyers' market. Markets "governed" by sellers were replaced by markets "governed" by buyers. When the future of demand becomes uncertain, and thus difficult to predict due to both the variability of the economic environment and the fragmentation of markets into several segments or niches, the risk of investments for companies increases, and many adopt a "wait and see" attitude, decide not to invest, or limit their investments.
- The introduction of new technologies and new production methods also contributed to reducing the stability of demand on the markets. Two examples suffice: (1) lean production, which allowed for lower costs at smaller volumes with higher quality standards in the automotive industry; and (2) computer-integrated manufacturing, which made it more economical to design and to obtain greater product variety.

"De-maturity". The effect that must influenced marketing strategies was the "de-maturity" of the industry, which increased the proliferation of products and paved the way for mass customization. The term "de-maturity" was coined by Abernathy, Clark and Kantrow in 1983. It meant that the automotive industry had entered a phase of sustained innovation, instead of suffering a decline in terms of product and process innovation as in those industries that had reached maturity.

During the 1980s and early 1990s, the automotive industry in fact displayed a high degree of product and process innovation. Among other things, product innovation brought the industry the all-wheel drive, air bags, microprocessors controlling more functions, and navigation systems such as the infrared night vision display. On the other hand, process innovation developed manufacturing automation technologies such as robotic welders and painters, not to mention the dissemination of the process innovations introduced by the Japanese (just-in-time and total quality management became very popular in Western countries, for instance).

2   The Shift Towards Mass Customization

The conditions of the automotive industry changed in less than a decade. The old rules of mass production based on stability and control of demand were no longer sustainable. In the place of homogeneous markets, standardized products, and long product life cycles, a new market structure emerged, made up of heterogeneous and fragmented markets, a variety of products, and shortening product life cycles. The carmakers' response was, simply put: "mass customization" (Pine and Gilmore 1997; Gardner and Piller 2009). Later on, from the early 1990s, the internet greatly increased the possibilities of mass customization. Carmakers could assemble components and modules requested at the last minute before delivery.

While the main goal of mass production was to develop, produce, and market automobiles at prices low enough "that nearly everyone can afford them", the main goal of mass customization was to develop, produce and market affordable automobiles and services so that "nearly everyone finds exactly what they want" (as quoted by Pine II 1993).[2]

Table 1, taken from Pine II (1993), compares the characteristics of mass production with those of mass customization. It summarises what has been examined above.

As demand for automobiles became unstable, a variety of fragmented segments and niches took the place of old, standardized products in the market. The negotiation power shifted from producers to buyers, who demanded products that would better respond to their expectations. The variety of products offered by carmakers required flexibility in the manufacturing process, and the production system thus had to be

Table 1 Mass customization versus mass production

	Mass production	Mass customization
Focus	Efficiency through stability and control	Variety and customization through flexibility and quick response
Goal	Developing, producing, marketing, and delivering goods and services at prices low enough that nearly everyone can afford them	Developing, producing, marketing, and delivering affordable goods and services with enough variety and customization that nearly everyone finds exactly what they want
Key features	• Stable demand	• Fragmented demand
	• Large, homogeneous markets	• Heterogeneous niches
	• Low cost, consistent quality, standardized goods and services	• Low cost, high quality, customized goods and services
	• Long product development cycles	• Short product development cycles
	• Long product life cycles	• Short product life cycle

Adapted from Pine II (1993), p. 47

_____

[2]Mass customization is a "production process that combines elements of mass production with those of bespoke (made to order) tailoring (designed to suite specific need). Products are adapted to meet a customer's individual needs, so no two items are the same". Mass customization uses some of the techniques of mass production. For example, its output is based on a small number of platforms, core components underpinning the product. In the case of a watch, the internal mechanism can be personalised through a wide variety of options at later stages of the production. The same is increasingly true for cars. Even a traditional mass production manufacturer like BMW boasts that no two cars are identical.

changed. Shorter development cycles and product cycles were needed, as well as general-purpose machinery and highly skilled workers.

Customization meant closely meeting consumer expectations and demands, often allowing producers to charge higher prices. The resulting higher profits could balance out any loss of efficiency due to the lower volume of each individual product due to demand fragmentation.

A set of new objectives. In mass customization, the objectives of companies' functions partly change. Production is directed at achieving efficiency in all processes, while in mass production priority is given to the production process. R&D is geared at incremental innovation, while previously it had sought to introduce radical innovations, such as Ford's assembly line and Budd's all-steel vehicle. Financial and accounting functions support the firm's strategies, providing them with information that is useful for long-term and short-term decisions, while in the "old competition" the main goal was to produce external financing reports. Changes in marketing functions were even more profound, as illustrated in Table 2. Marketing functions changed drastically (Webster 1988).

Table 2 offers a succinct comparison of the evolution of marketing functions from the "old competition" to the "new competition". This is followed by a commentary on the detrimental effects of the "old competition" as opposed to the positive effects of the "new competition", as identified by Pine II (1993).

The detrimental effects of the "old competition". In the "old competition", selling to homogeneous markets, large volumes of low-cost standardized products and

Table 2   The "old competition" versus the "new competition": marketing functions

The "old competition"	The "new competition"
Focus	Focus
• Selling low-cost, standardized products to large, homogeneous markets	• Gaining market share by fulfilling customers' wants and needs – first domestically, then in export markets
Primary benefit	*Detrimental effects*
• Stable, predictable demand	• Too "enamoured" of technology
Detrimental effects	*Positive effects*
• Disregard for many customer needs and wants	• Ability to respond quickly to changing customer needs
• Disgruntled, disloyal customers	
• Opening up of market niches	• Filling the niches
• Segment retreat and avoidance	• Market takeover
• Lack of exports	• High sales, domestically and through exports
	• Technology-intensive products

Adapted from Pine II (1993), p. 128

using marketing as a leverage resulted in more predictable demand, but many detrimental effects started to display their consequences.

- Selling higher volumes of standardized products to homogeneous markets only allows for lower prices, but the natural consequence of ignoring what is not included in the standard is refusing to recognize a part of customers' needs and wants. In this way, a firm might lose a highly significant share of the market, leaving it open for competitors. Conversely, the "new competition" tries to gain market shares by filling as many as niches as possible. This was the case with the choices that led to Henry Ford's demise. In the 1920s, as a result of mass production the prices of the Model T continued to drop, but consumers had changed. Low prices were no longer enough to convince them to buy. Ford left GM part of the market, which wanted innovation was willing to pay the right price to get it. By contrast, GM was able to quickly respond to changing customer needs.
- In the "old competition", the goal was to mainly sell what carmakers had already built. If low costs were not enough to keep demand stable, carmakers used massive advertising and price promotions to sell. Once more, they gave up on understanding and responding to customers' needs and wants. But if customers were persuaded to buy what they did not really want, their dissatisfaction with the product soon became clear. In the end, that led to disappointed customers and a lack of loyalty.
- While mass producers refused to change their policy of standardized products to be sold to homogeneous markets, more flexible competitors were able to fragment the full market into an increasing number of niches and to fill those niches with products that better met customers' expectations. Often, opening up market niches to new competitors later paved the way for a dominant position, as the Japanese had managed to do in the U.S. and Europe in some segments of the automotive market.
- When a new competitor enters an industry, it starts serving a segment that its incumbents have ignored or only partially served, usually at the low end of the market with lower profit margins. The newcomer's long-term goal is to later expand their position, introducing innovations and a variety of products. Mass producers are inclined to retreat from those segments and to avoid competition because they estimate that the level of profit margins are not such as to justify resisting the attack. There are many examples of newcomers that have conquered segment after segment until they occupy a strong position in the market facing limited opposition from incumbents. Sometimes, incumbents wanted to resist but had a limited ability to do so. In the U.S., Japanese carmakers now held a dominant position in a segment that the Big Three had purposefully disregarded. Hyundai and Kia followed the same course of action in Western Europe.
- Having chosen the strategy of selling large volumes of low-cost, standardised products on homogeneous markets, manufacturers gave up on selling beyond the national borders, where adaptations were generally required to meet the requirements of local demand. One of the consequences of this, often the main one, was that demand became more unstable because manufacturers depended solely on the market of origin. Another disadvantage was that the company's marketing lost the

chance to gain experience by dealing with the traditions, market structures, and requirements of different consumers.

The positive effects of the "new competition". As Table 2 shows, Pine II (1993) opposed the "detrimental effects" of the "old competition", one by one, with a set of positive effects: (1) the "ability to respond quickly to changing customer needs"; (2) "filling the niches"; (3) "market takeover"; (4) "high sales domestically and through exports"; and (5) "technology-intensive products". Let us briefly examine those effects.

1. When they introduce a new product, carmakers are not always certain they have identified which niche it will occupy, inasmuch as consumer purchasing behaviour is constantly changing, as are the marketing strategies of the competition. If the consumer's response is not as anticipated, carmakers must quickly correct the offer. Often, they introduce new products, or modified products, based on the indications of consumers' responses. This all entails quick responsiveness. Given the size of the investments, in the automotive industry listening to the "voice of the consumer" is very important.
2. In the "old competition", carmakers did not want to cover the entire market. As they were seeking stable, homogeneous demand, they could define their target marketing only through those two characteristics. In such a way, they left the way open for new competitors that could secure their market share by filling the niches. To be successful, they needed to understand the customer and be flexible in offering a variety of products.
3. Companies that embrace mass customization never not stop looking for niches. They move from those they already occupy to adjacent niches. They are continuously looking to create new niches. They fragment the broader marketplace by seeking to open up new niches that satisfy new needs and wants. Carmakers in the "old competition" succeeded in a different way. When a new competitor entered an industry, it would start serving a segment that its incumbents had ignored or only partially served. Mass producers were inclined to retreat from those segments and avoid competition because they estimated that the level of profit margins did not merit any resistance.
4. While in the "old competition" carmakers gave up opportunities to export, thereby missing out on the chance to understand different cultures, different customers' needs and how to respond to them through a greater variety of products, in the "new competition" the strategy was reversed. Management realised that gaining a market share in export markets takes time and may generate little or no profit for years, but that in the long term offering a variety of products suited to local demand can reap rewards.
5. Competing in a large variety of segments and niches is a powerful stimulus for honing R&D in the search for incremental innovation. The more product variety is increased, the more intense technological research becomes, the more technology-intensive products are released onto the market, and the more new areas there are to be explored, as attested to by the advances in "no-hands drive".

According to Pine II (1993), after launching the mass customization strategy, the objective of lowering the production costs of individual products and services is pursued through five primary methods: (a) creating products and services that can be adapted to the needs of the client (customizable); (b) integrating standardised products with customised services before they are sent to customers; (c) offering customization in points of delivery, thereby also guaranteeing customization in dealerships; (d) providing a quick response to customers' needs and wants throughout the whole value chain; and (e) modularizing components to make customizing end products and services easier.

3   Understanding People and Cars

As well as the proliferation of models and market saturation, in the richest countries in the 1970s marketing also began to be influenced by consumer behaviour concerning mode of transport and vehicle choices. Various studies have examined these aspects in depth and sought to ascertain which factors influenced such choices the most (Kohler 2006; Newman 2013; Small 2013). Whitmarsh and Xenias (2015) have distinguished two types of factors that affect buyers' car choices: (1) factors that affect choices about modes of transport, thus how to travel; and (2) factors that influence vehicle choice. Directed at understanding consumer behaviour, these studies partly altered the marketing approach. Previously, psychography and demography had been the main referents for market segmentation and market targeting.

(1) Mode of transport choice. Various factors influence mode of transport choice. Some are more important than others. Much depends on the context: for example, in rural areas where public transportation is scarce, the automobile is for many irreplaceable, while in urban areas several alternative modes of transport exist.

   Whitmarsh and Xenias have observed a "behavioural lock-in" in the use of automobiles, which generates considerable resistance to changing attitudes. The automobile is the primary form of personal transportation in richer societies, and the structure of cities has been constructed around road networks and automotive traffic requirements. People who are accustomed to driving do not seriously consider other modes of transport even when they are better; on the contrary, they tend to exaggerate the poor quality of alternative modes. Consequently, predicting choices when it comes to the modes of transport choices is somewhat complex. The two authors conclude that, "Model choice is often shaped by unconscious habit and available infrastructure".

(2) Vehicle choice. Vehicle ownership can be predicted based on income and other socioeconomic factors, including age, the size of the family unit, and the number of people driving the car. The lifestyle and personality of the owner also have an impact. From a psychological perspective, the choice is also influenced by non-rational economic factors. Owners often underestimate running costs and

overlook fixed costs. Given the high price they have paid, they feel obliged to use the car frequently. Cars are often a symbol of "conspicuous consumption". This concept was first described by Veblen in 1953, when he observed that for certain products, the laws of demand are not applied, but rather overturned. Certain people buy a car because it costs a lot. Their goal is to use the car to convey their wealth and power. As a consequence, vehicle choice often overlooks economic factors and is often "shaped by social factors such as identity, status, and interpersonal communication".

References

Abernathy W, Clark K, Kantrow A (1983) Industrial renaissance: producing a competitive future for America. Basic Books, New York

Gardner D, Piller F (2009) Mass customization. Happy About, California

Kohler J (2006) Transport and the environment: the need for policy for long term radical change. In: IEE Proceedings intelligent transport systems 153(4):292–301

Newman D (2013) Cars and consumption. Capital Class 37(3):454–473

Pine II J (1993) Mass customisation. The new frontier in business competition. Harvard Business School Press

Pine B, Gilmore J (1997) The four faces of mass customisation. Harvard Bus Rev 75(1):91–101

Small K (2013) Urban transportation economics. Routledge, London

Veblen T (1953) The theory of the leisure class. Mentor Books

Webster F (1988) Rediscovering of the marketing concept. Business Horizons 31(3):29–39

Whitmarsh L, Xenias D (2015) Understanding people and cars. In: Nieuwenhuis P, Wells P (eds) The global automotive industry. Wiley

Chapter 14
Braced for a New Model of Creating Value

Abstract In the mid-1990s, in Western markets, the evolution of industries and technologies was accompanied by an emerging shift in competitive advantages from "upstream" activities towards "downstream" ones. Info-intermediaries emerged, such as Autobytel.com, which took power away from carmakers (OEMs) in distribution and granted more power to consumers. The centre of gravity shifted towards the "downstream" area, and therefore also towards marketing, requiring many principles to be revised. This also paved the way for the entry of new competitors onto the market. The car industry resisted these threats. It was more resilient to the value migration than expected, but for the management of major carmakers it was clear that the change called for a response to important strategic questions such as: "Where will profits emerge in the new digital infrastructure?" and "How can we reshape our business to take advantage of the new opportunities better and faster than competitors?". The responses to these questions all involved the incorporation and assimilation of new digital technologies. On the threshold of the early years of the new millennium, the advances towards a transformation in the automotive industry were already clear. Information technology had begun to be deeply integrated as a tool in marketing research, and in the other main functions: from sourcing to product design, from logistics to manufacturing, and from marketing to after-sales services. Products could be better designed and manufactured. Customers' reactions and expectations could be quickly understood and analysed. Cars became more reliable, maintenance less frequent, and repairs rarer.

Towards the mid-1990s, the effects of advances in digital technology began to come to light in the car industry too. Info-intermediaries emerged, such as Autobytel.com, which took power away from carmakers (OEM) in distribution, and gave more power to consumers. Once again, marketing needed to adapt itself to change. Carmakers largely managed to preserve their position of strength by investing in the "core" of the product—power train and transmission—and brand loyalty, as well as by using new technologies to better understand the expectations of prospective consumers. However, the transformation initiated by the digital age was unstoppable.

© Springer Nature Switzerland AG 2019

E. Candelo, *Marketing Innovations in the Automotive Industry*, International Series in Advanced Management Studies, https://doi.org/10.1007/978-3-030-15999-3_14

Autobytel.com. Based in Irvine, California. It was founded in 1995 by a former car dealer, Peter Ellis. The company site provided car buyers with a large amount of information about car models and pricing. At the beginning of the new millennium, it operated with around 3000 North American car dealers, which paid monthly fees to belong to its network. When a consumer interested in purchasing a car logged on to www.autobytel.com, they could check many features of a vehicle on offer: from prices to financing options, and from types of insurance to reviewers' remarks. If the consumer did not have a specific model in mind, they could indicate a price range and their desired features. In response, Autobytel listed the possible options available for the purchase on the screen. After the consumer had identified the various elements of their choice, on request Autobytel indicated in which dealership the vehicle was ready for delivery. Within twenty-four hours, the dealer affiliated with the network had to make contact with the client and send an offer with a no-haggle quote. J.D. Power estimated that at the end of the 1990s, Autobytel would be the leading e-tailer of the car industry, with 45% of all new vehicles sold over the Internet (Boulton et al. 2000).

1 Reality Has Started to Bite

The change shook up marketing in the industry. Like other info-intermediaries in the car industry (e.g. Autoweb), Autobytel emerged on the web and took control of customer relationships. In *Unbundling the Corporation* (2000), Hagel III and Singer observed that Autobytel benefitted both parties, both prospective customers and dealers. The site (Autobytel) gathered information on prospective customers and their preferences, selected this information and put prospective customers interested in making a purchase in contact with dealers that could make an offer that met their requests.

As a consequence, the role of the traditional car dealer inevitably changed and car companies had to rethink their marketing strategies. The new info-intermediaries:

- took the acquisition and management of customer relationships away from traditional dealers. As such, dealers lost part of their customer relationship business both before and after the purchase. First and foremost, they remained supervisors of the showroom and after-sales services.
- As they acquired a lot of information provided by potential customers, they could better understand the motivations that drove them to purchase a certain model of car.
- As such, they could also assist customers in choosing financing options and the best insurance package. They could make a list of the best repair shops, the best rescue services and suppliers of other services. They could also take responsibility

for reminding customers when periodic maintenance and overhauls were due. Hagel III and Singer (1999) demonstrated that car manufacturers could also supply these services to anyone purchasing their products, but they did not have all the information that info-intermediaries could collect in the pre-sales stage. They therefore did not have all the information on why one of their products had been chosen over those of their competitors (Carr 1999).

The transformation dated long back. In 1989, while the Internet was undergoing a rapid diffusion, Tim Berners, a researcher at CERN in Geneva, came up with the idea of linking contents stored on different computers. This was a breaking point because up to that point the Internet had been a tool used by a community of technologists and scientists. When the first web page on the Internet was created in 1991 at CERN, a large audience of people and organizations had the tool they needed to directly exchange information. The development quickly became exponential. From that point on, the Internet, the web and digital media slowly but inexorably transformed marketing and business, especially in the car industry.

The Internet offered customers a large choice of products and services, a means to choose what to buy more quickly, and at the best prices. For company marketing, it made it possible to enter new markets and present new products and, above all, it offered small companies the opportunity to rival the largest ones. Not only did it give marketing in the automotive industry the capacity to significantly expand the circle of people it could reach, but it also made it possible to restrict the target (prospective customers) to be reached by making communications more precise and efficient.

In *Internet Marketing* (2000), Chaffey et al. underscored that, at the beginning of the new millennium, marketing in many companies was starting to reflect on whether, with the arrival of the Internet, the concepts, theories and tools of traditional marketing were still valid, and what might change in distribution channels, communication and brand loyalty plans.

Concerning the effects of the Internet on marketing management, the obvious response was that much depended on the characteristics of the company. Very strong effects could be envisaged in companies based on electronics, less strong ones in large distribution companies (i.e. Unilever), and in companies in which the physical content—"atoms"—of their products was significant (i.e. naval and aeronautical constructions, cars, and mechanical constructions in general). In marketing management, different effects could be foreseen depending on the elements of the marketing mix; these effects were certainly considerable in relation to *place* and *promotion* given the irruption of the ubiquitous Internet. These effects could not be ignored when it came to pricing and products. The Internet of Things was on the verge of appearing.

In the automotive industry, consumers increasingly frequently went online to research, make comparisons, evaluate, and make their choices. They based their choices on the quality of the offer compared to their expectations. They then moved to bricks-and-mortar by making contact with the dealer to negotiate and seal the transaction.

2   The Shift Toward "Downstream"

Another factor also fed the change. Towards the end of the 1990s, on Western markets, with the evolution of the industries and technologies, the movement in competitive advantages from "upstream" activities towards "downstream" ones began to occur (Tapscott et al. 2000; Jacobides and MacDuffie 2013).

For decades, competitive advantages had been sought and achieved primarily in "upstream" activities: how to seek production factors at lower costs; how to bring about greater efficiency in production processes; how to construct larger facilities in the search for larger economies of scale; and how to devise and develop new products. However, these methods had been largely imitated, and the related advantages had gradually and inexorably disappeared. The centre of gravity had shifted towards the "downstream" area, and thus also towards marketing, requiring the revision of many principles.

On the threshold of the new millennium, *"the sources and locus of competitive advantage now lies outside the firm..."* wrote Dawar in an article that came out in *Harvard Business Review*, with the title 'When Marketing is Strategy'. This was also true in the car industry, where sourcing from the best first-tier suppliers and belonging to the most rewarding alliances often distinguished companies from their competitors. Moreover, marketing choices became more important as it was no longer sufficient to have the best products; it was necessary to select the most appropriate target and position products on the market in the most efficient way compared to competitors. This was because the change in markets was increasingly dictated by changes in customers' purchasing criteria rather than by substantial product improvements.

3   Standing up to the Winds of Change

Technological advances had a disruptive effect and value was no longer stable in the various phases of the value chain, as it had been in the past, beginning to migrate upstream or downstream in many industries. This led to greater uncertainty in strategies and paved the way for the entry of new competitors onto the market.

The car industry had resisted this threat. It was more resilient to value migration than expected. Contrary to the opinion of many, the major carmakers kept control of the value chain. In 1999, authoritative experts had predicted the same fate for the car industry as for personal computers, where the incumbents, such as IBM, lost power and value (capitalisation and profits) migrated towards the suppliers Microsoft and Intel, in particular. *"In the 1999 report 'The dawn of mega supplier', Bain & Company predicted that new giant suppliers would achieve pre-eminence in the auto industry by designing vehicle systems that could be standardized within and across OEMs"*.[1]

---

[1] Quoted in Jacobides and MacDuffie (2013).

According to this prediction, there would be a major movement in profits in the car industry towards a few suppliers of standardised forms and components. The leading manufacturers, such as GM, Ford, Daimler, BMW and Toyota, would be confined "*to simply assembling and marketing cars made up of those components*", as had occurred in the computer industry. Although they largely resorted (and continue to resort) to outsourcing (which makes up around 70–75% of the final price), and although competition between incumbents was (and still is) extremely strong, for the most part the major carmakers maintained their relative share of market capitalisation and profits. Value primarily remained with the OEMs and did not migrate upstream (towards the suppliers) or downstream (towards aftermarket products and services). Of course, first-tier suppliers such as Bosch, Continental and TWR firmly acquired leading positions in the supply chain, but modestly impaired the power of the major carmakers.

As Jacobides and MacDuffie (2013) observed, carmakers and other major firms like Apple and Google were able to hold onto their control of the value chain in three main ways: "controlling the assets least likely to be commoditized"; "serving as a guarantor of quality to end customers"; and "staying in close touch with changing customer needs" (Jacobides and MacDuffie 2013).

1. "Controlling the assets least likely to be commoditized". The OEMs successfully maintained a strong capacity as "system integrators", that is, the capacity to assemble components and modules in the end product. This is attested to by the fact that in the past forty years, no new entrant has managed to stand up against the main manufacturers in the industry. The last, in order of time, was the Korean company Hyundai. Magna (Austria) and Valnet (Finland) did not go beyond assembling individual car models on behalf of the main manufacturers, including Daimler and BMW. Indian and Chinese manufacturers failed to beat the competition presented by joint ventures between foreign companies and other companies from their own nation. They chose the strategy of acquiring historic brands in the premium and luxury segments, such as Tata (India), with the acquisition of Jaguar and Land Rover, and Geely (China) with the acquisition of Volvo. Tesla successfully gained more entry into the industry, but for years the electric car remained a market niche.

   How did the major players manage to maintain control? (a) Above all, in car production, the integration of component parts and modules requires much more complex technology than those for assembling PCs and other products (whose companies had suffered the disruptive effects of new technologies); (b) carmakers made major investments in power units and transmissions that were (and still are) destined to remain the "core" of the vehicle for a long time; (c) they avoided depending on a single supplier, even at the cost of maintaining or reviving forms of vertical integration.

2. "Serving as a guarantor of quality to end customers". Marketing responded firmly to new technologies. Car manufacturers fiercely protected their image as final producers, excluding any label that called to mind the brand of a supplier from vehicles as much as possible, to such an extent that few clients would know who

the supplier of their vehicle was, even when the latter provide extremely important components such as ABS, EPS and airbags. The tyre brand is one of the few
exceptions to this rule. This was not the case in the PC industry, where Microsoft
and Intel managed to acquire far greater recognition (and market capitalisation)
than that of assemblers (IBM was forced to hand over the PC division to the Chinese company Lenovo). In the automotive industry, this was especially true for
premium manufacturers such as Audi, BMW and Mercedes. While presenting
themselves as the sole guarantors of the final quality of the product, car manufacturers also assumed liability when the client complained about the vehicle
malfunctioning (unless they made a claim against the supplier if this was due to
a specific component). This assumption of responsibility represents a possible
cost, but it provides a strong competitive advantage in terms of marketing.

3. "Staying in close touch with changing customer needs". In the car industry,
    the final consumer's needs did not change substantially in those years: personal
    mobility; fuel efficiency; functional and emotional attributes. In the early 2000s,
    the threats presented by the electric car (Egbue and Long 2012), and the driverless
    car were still far-off. In the history of the car industry, we have many examples
    of successes owing to new products introduced into the market to interpret new
    requirements, such as Minivan, SUV, and crossover.

4   In Search of the Winning Formula

The threat of a migration of value upstream (benefiting certain suppliers) and downstream (towards networks of dealers and consumers) had been stemmed, but for the
management of major carmakers it was clear that the change called for a response
to strategically important questions such as: "Where will profits emerge in the new
digital infrastructure?"; "Who will take hold of them?"; and "How can we reshape
our business to take advantage of the new opportunities better and faster than competitors?" (Carr 1999).

The cases of AutobyTel and Autoweb.com had shown that the effects in the
automotive industry were certainly far from negligible.

At the same time, carmakers' marketing management tried to grasp the opportunities deriving from the advances in digital technology to manage relationships
with customers to their own advantage. For instance, the diffusion of the Internet
convinced GM and other carmakers that their future was not only to transform themselves from "carmakers" but also to become a "communications intermediary". They
were tempted to provide information to restaurants, service stations and retailers that
would have paid to know when a car was passing nearby. They decided, however,
that the time was not yet right to fully take advantage of this opportunity.'

After all drivers spend an average of 8.5 h a week inside the approximately 70 million cars GM vehicles on the road today. By comparison, American Online's 22 million subscribers spend 7.5 h a week online. The information in every one of GM's vehicle is immensely valuable to other marketers. Shell and Texaco, for instance, would pay good money to know how much gas is left in a car tank. Retailers and restaurateurs would pay to know when a vehicle is passing nearby. Merchants would pay for access to a GM vehicle's service history. By reconceiving the car as an information device, GM dramatically increases the amount of value it can capture from each vehicle, while providing services that tie car owners closer to the company. (Kenny and Marshall 2000)

On the threshold of the first years of the new millennium, the advances towards a transformation of the automotive industry were evident. Information technology had begun to be deeply integrated as a tool in marketing research, and in the other main functions: from sourcing to product design, from logistics to manufacturing, and from marketing to after-sales services. Products could be better designed and manufactured. Customers' expectations and reactions could be rapidly understood and analysed. Cars became more reliable, maintenance work less frequent, and repairs rarer. There were problems, however.

In the 1990s, the cost of car software and electronic parts had increased. Two effects were destined to reproduce themselves, even successively. Firstly, the technology of the electronic part and the software enjoyed an exponential rate of development (far superior to that of the mechanical parts). Over time, electronic components and software became exponentially better and cheaper than mechanical parts. Clients who experienced this trend in smartphones and other digital devices expected the same trend/response in cars. And secondly, carmakers therefore needed to choose between lowering the price or maintaining it, while continuously enriching the technological content. Most manufacturers chose the second option, triggering a spiral that not only rendered competition more intense, but also meant that the average customer used and appreciated only a part of the offer.

5 Identifying Benchmark Competitors

In the second half of the 1990s, advances in digital technology led to competitor analysis and competitive position acquiring an important role in the marketing strategies of carmakers. As had already occurred on several occasions, despite the vast opportunities offered by technology, competitor analysis nonetheless suffered from the restrictions dictated both by limited resources and by the need to focus on clearly defined market areas through the most precise targeting possible. It quickly became clear that marketing management could not examine the competitive capacities of each competitor. Thus, a technique was needed to delimit the analysis and make the use of resources more efficient.

In *Market based management* (1997), Best suggested building a "perceptual map". Starting from the principle that the closer a prospective customer saw the offers of two manufacturers as being, the more likely they would be to go from one to the other, while the further they saw a pair of competitors as being from one another, the less likely they would be to do so. Moreover, the potential customer could assign a weight (a rate) to the differences/distance noted between a competitor's offer and that of their ideal product or supplier. Based on these perceptions and weights (rates), a "perceptual map" could be created, which is very useful for understanding the competitive position of one's offer and identifying the competitor to be considered as a benchmark.

Best presented an example, summarised below, of a perceptual map useful for identifying which competitor(s) to identify as a benchmark.

Luxury car market. In the mid-1990s, in the luxury car market in the U.S., Volvo 700, Mercedes 420, BMW 525, Lincoln Towncar, Buick Regal, and Honda Prelude were seen as competitors. Volvo 700 and BMW 525 were deemed very close in "perceived similarities", while Lincoln Towncar and Honda Prelude were rated as very dissimilar. Prospective customers were asked to assess the differences between the six brands and their ideal car. The assessment produced two different segments, identified as segment A and segment B, having different sets of customer needs and product differences. "The ideal car for segment A" was almost equidistant from the Volvo 700, BMW 525, Buick Regal, and Honda Prelude. As such, in segment A, the client's probable choice could be limited to these four brands. The marketing of Buick Regal therefore needed to identify Volvo, BMW and Honda as key competitors in serving segment A, while Mercedes 420 and Lincoln Towncar were equally close to the Buick Regal. If, on the other hand, Buick Regal was more interested in serving segment B, then Mercedes 420 and Lincoln Towncar would have been their competitors to benchmark.

The inter-brand differences can be represented graphically on a two-dimensional map, which helps identify which rivals the firm is competing with in a given segment, and what the company's competitive position is compared to its rivals in attracting prospective customers.

References

Best R (1997) Market based management. Prentice Hall
Boulton R, Libert B, Samek S (2000) Cracking the value code. How successful businesses are creating wealth in the new economy. HarperBusiness
Carr G (1999) (ed) The digital enterprise. Harvard Business Review Book
Dawar N (2013) When marketing is strategy. Harvard Bus Rev 91(12):100–108

Egbue O, Long S (2012) Barriers to widespread adoption of electric vehicles: an analysis of consumer attitudes and perceptions. Energy Policy 48:717–729

Hagel III J, Singer M (1999) Unbundling the corporation. In: Carr G (ed) The digital enterprise. Harvard Business Review Book

Jacobides M, MacDuffie J (2013) How to drive value your way. Harvard Bus Rev 91:92–100

Kenny D, Marshall J (2000) Contextual marketing. The real business of Internet. In: Carr G (ed) The digital enterprise. Harvard Business Review Book

Tapscott D, Ticoll D, Lowy A (2000) Digital capital. Harnessing the power of business web. Harvard Business School Press

Part IV
The Digital Age: The Changing Face of Marketing

Chapter 15
Is Disruption Taking Apart
the Carmakers' World?

Abstract The word disruption, in its original sense, expresses the concept that even successful companies can fail, despite continuing to do what they did before well. In the digital era, it is often used both when a company is well-managed but fails, and when it fails because it is badly managed. In the car industry potential disruption is generated by the combined effect of more digital technologies, platforms (Uber, Lyft, Didi), electromobility (Tesla), and autonomous driving (Apple and Google). Disruptors in automotive industry are big data and machine intelligence companies, at their core; they use direct-to-consumer models; they are venture-backed startups; they make decisions with incomplete data and uncertainty; they hire the best; they are organised in small teams; and they take risks and break rules. Choosing the most appropriate moment to enter the market or drastically change strategies is more important than ever in a highly changing environment. The history of technological innovation is full of potential disruptors that could offer superior performance to rivals, but which failed to identify the right moment to enter the market. Finally, how can disruption be managed? The first step is to predict the impact of a new disruptive business model on consumer behaviour. The second involves estimating how many consumers could move to the new product. In the automotive industry, how many consumers would adopt a level 4 or level 5 driverless car? The third step is to extend the analysis of the impact of a disruptive business model to other related sectors.

The word disruption is frequently associated with digital transformation. In its original sense, disruption expresses the concept that even successful companies can fail, despite continuing to do what they did before well. Now, though, it is a word that is often used to indicate situations that differ greatly among themselves, to the point that it has lost a lot of its usefulness. It is used both when a company is well-managed but fails, and when it fails because it is badly managed.

The following pages sum up the development of this concept, review what five famous disrupters of the automotive industry do, what their actions have in common, what the sources of disruption are, and how firms can manage it.

Following the different theories and looking at past experiences is not an academic exercise; rather, it helps understand where disruption comes from, the effects it can have on marketing, and how it has been managed thus far.

© Springer Nature Switzerland AG 2019

E. Candelo, *Marketing Innovations in the Automotive Industry*, International Series in Advanced Management Studies, https://doi.org/10.1007/978-3-030-15999-3_15

1   The Forerunner

In *Capitalism, Socialism and Democracy,* published in 1942, Joseph Schumpeter describes the phenomenon of "creative disruption", through which innovation destroys old companies and economic systems after creating them. Schumpeter considered the destruction of what it had previously generated to be endemic to capitalism. He did not use the word disruption, but this concept can be considered as part of a line of descent from that of "creative destruction".

The concept of disruption reached management's attention in 1995 with an article by Christensen and his subsequent book *The Innovator's Dream* (Christensen 1997), where he observed that companies can fail precisely because they are well-managed and have been successful. Many authors followed his path. With *The Disruption Dilemma,* Gans stands out for having helped clarify the many aspects of the concept of disruption, and for identifying where disruption can come from and how its effects can be managed. For Gans, "the phenomenon of disruption occurs when successful firms fail because they continue to make the choices that drove their success", and he specifies that a company can fail because it is badly managed, but this did not necessarily occur as a result of what is understood by disruption. Gans also draws a distinction between the three types of disruption deriving from a new technology and those caused by the arrival of a new competitor. The current case in the car industry is a potential disruption generated by the combined effect of the progress of more digital technologies, the use of platforms (Uber, Lyft, Didi, and others), electromobility (Tesla, precursor), and autonomous driving (Apple and Google).

Companies or entire sectors that fall victim to disruption will not necessarily disappear. The self-driving car pioneered by Google might become commonplace, the main mode of transport within two-three decades.

This could revolutionise transport, but the traditional car is destined to coexist (certainly in emerging markets). Disruption is underpinned by the idea that the new company satisfies the clients' requirements with products, services, or a business model with characteristics the incumbents cannot provide.

2   Sources of Disruption

Like others that have discussed the concept of disruption, Gans considers the case of the Encyclopedia Britannica, which illustrates all the elements of disruption and how it can be avoided. He then considers the case of Blockbuster, which succumbed because it did not know how to react to the introduction of a new business model. Their analysis is useful for interpreting what is happening in the car industry, and how companies could respond.

- Disrupted by a new technology. Encyclopedia Britannica (E.B.) waned, but then resurged. E.B. sold encyclopedias directly, door to door, at extremely high prices, with great success. Clients bought them thinking that they were making an investment in culture for themselves and their children. The product was very high-quality, written by experts from various sectors, including many Nobel laureates. E.B. took a downturn when Microsoft launched Encarta, an encyclopedia for which millions of CDs were sold. E.B. responded by launching the CD-ROM version of its famous encyclopedia, but at a price too far above that of Microsoft and other competitors who had entered the market in the meantime. In its turn, Encarta suffered the same fate as E.B. as a result of the unexpected success of Wikipedia. After various attempts, E.B. decided to give up on the printed version of its product and to dismiss its sales staff. It then managed to recover by harnessing the new technology that transformed the industry in which it operated. The new technology was the computer, which was gradually making its way into homes. For this reason, those commenting on the affair observed that E.B. (in the printed version) was not fighting a losing battle with Encarta but rather with the computer.

- Disrupted by a new model. Blockbuster, on the other hand, succumbed to a new business model. Blockbuster filed for bankruptcy in 2010 after having dominated the bricks and mortar rental market of DVDs and videocassettes for two decades. At the peak of expansion, it had 9000 stores spread across the world. It failed because a new competitor arrived: Netflix, which introduced a new business model, interpreting the use of existing technology—the DVD—in a new way. These new products are lighter than videocassettes. Netflix sent them out by post, meaning that the client did not have to go to the Blockbuster store to take them out and go back to return them. Netflix entered the market slowly. It attracted only a small proportion of the potential Blockbuster clients. The latter responded by imitating Netflix, but it did not focus on postal deliveries despite having ample resources because, by imitating the new entrant completely, it was afraid of diluting its leadership image and losing what distinguished it from its new rival. When the internet arrived a few years later, Netflix harnessed the new technology, moving its customers over to video streaming. Blockbusters failed not because of a new technology, but because of a new business model that attracted consumers in a new way.

The theory proposed by Christensen explained why a company that continues to serve traditional clients, ignoring new technologies, can be inexorably dragged down into bankruptcy. There are many examples of companies, even in the automotive industry, that have gone bankrupt because they continued to do what had brought them success.

Christensen's theory can be summed up thus: A new entrant in the industry introduces a product innovation in the low price range, making it accessible to those that cannot afford what the incumbents are offering. The new product is inferior in terms of performance and features, and therefore the incumbents do not consider it as a threat. Having acquired a good understanding of the market and buyers' expectations, the new entrant begins to constantly improve upon the product, maintaining the low price until the product is "good enough" for the incumbents' traditional clients. Those that had ignored or underestimated the danger cannot compete with the new arrival and often go bankrupt or are forced to leave the market.

To support his theory, Christensen only offered examples of B2B sectors, such as mechanical excavators, in which buyers are not consumers but businesses, and this raised criticism. Christensen responded to this with a successive publication, *The Innovator Solution*, distinguishing between two sources of disruptive innovation—the low-end and new market—but without convincing his opponents.[1] When Christensen was interviewed about the introduction of Apple's iPhone, he said that it would not be able to disrupt the incumbent mobile phone manufacturers like Nokia. Subsequently, after the resounding success of the iPhone, Christensen argued that the iPhone was indeed a disrupter, but of the personal computer industry rather than Nokia.

Gans (2017) moved beyond these positions by distinguishing between demand-side disruption and supply-side disruption, two forms that can act in a combined manner, as is the case in the automotive industry.

Demand-side disruption. This occurs when companies that are customer-focused continue to pay attention to their clients' demands, ignoring the arrival of other companies that began by serving the clients overlooked by the incumbents. Also making use of a new technology, the new entrants quickly become serious competitors. The incumbents go bankrupt or find themselves in serious difficulty because they do not react in time to recover before the new entrants gain dominant positions on the market. They do not understand, or do not want to understand the danger because the new entrants offer a low product or focus on unattractive segments.

Uber and Lyft and the likes have used the new technology of platforms to disrupt the taxi service protected by licenses by offering smartphone-hailed rides in major metropolises with anyone wishing to be a driver to transport other people. Initially, they only met with resistance from taxi drivers. Harnessing the new technologies, the new entrants created value for clients (low transport price with ride-hailing) and weakened the positions of traditional manufacturers. Although the new technologies eroded the profits of the incumbents, they were unable to immediately compete with a similar offer. Other incumbents—i.e., Daimler and BMW—also decided to enter the new market, with the goal of understanding, but the pioneers had already gained a significant share and investors had already recognised their high prices on the stock market (for Uber, in particular).

Incumbents were also slow to grasp that new tools are required to understand how new consumer demands evolve. The automotive industry is changing, and so is its culture. Disrupters understood far sooner that it is necessary to study how people move around, not only with cars, and how people interact with technology; Google, Apple and Microsoft, for example, are some of the largest employers of anthropologists in the world.[2]

---

[1] For example, Thompson B, "What Clayton Christensen got wrong" *Stratechery*, September 22, 2013.

[2] For more than a century, the auto industry's cultural identity was defined by calloused hands and brute force shaping steel. But as the industry perches on the edge of a drastic technological

Supply-side disruption. Incumbents have difficulty adopting a new technology and confronting new competition because they fail to efficiently organise both the product configuration and its supply chain differently from the new entrants. This is the case for car manufacturers, particularly in relation to autonomous driving and the transition to EV.[3]

To introduce new technologies, it is necessary to configure the product differently from the traditional approach used by manufacturers for over a century. The structure of the supply chain and its players are just as different. Here, too, Google and Apple have an advantage. They are not weighed down by specialised assets and numerous employees. For example, in autonomous driving, they concentrate on software technology and mapping, potentially offering a service that uses vehicles produced by traditional manufacturers. One example is Fiat Chrysler's alliance with Google's Waymo, where FCA supplies its Chrysler Pacifica minivan and Waymo automates it.

Disrupters are flexible. They do not need to manage the transitions from old to new. On the contrary, incumbents must manage the transition from car manufacturer to provider of new means of mobility and new services. This is a change in business model. During the transition, the profits, or a portion of them, generated by the sale of traditional vehicles will be used to finance new ventures, which will likely result in further reducing the profitability of the automotive industry, already one of the lowest of the various industrial sectors. How will investors react?

The problem has been clearly stated by Reithofer, Chairman of BMW, who in an interview asked rhetorically, "What would happen to the group's powertrain plants if the company ceased building combustion engines?". We cannot say—he added—that tomorrow morning we will only be building electric cars. "The question is how we can manage the transition from relying solely on combustion engines for profits, since not one carmaker currently earns money selling electric cars".[4]

---

disruption, its mechanical engineering prowess is no longer enough, experts say. The autonomous and connected future is going to require a new set of tools. "Why the auto industry needs a new toolbox". *Automotive News*, 4 December 2017.

[3] In their article, Henderson and Clark have explicitly related the success or failure of incumbent firms to whether they needed to undertake in architectural and component innovation. Their theory is that incumbent firms are great at supplying component innovations but may be unable to adopt architectural innovation Henderson R., Clark K., "Architectural innovation: The reconfiguration of Existing Product Technologies and the Failure of Established Firms" *Administrative Science Quarterly*, 35, no. 1 (1990) 13.

[4] "As BMW marks its 100-year anniversary, it finds itself entering a new disruptive digital age commonly called the Fourth Industrial Revolution. A resurgent Mercedes-Benz, slowing China car sales and rising competition from tech juggernauts Google and Apple are just some of the challenges it faces. "BMW's innovation, adaptability will be tested in the 21st century". BMW is already changing as it prepares for a time when it is no longer on top. "Volume is not everything" said Reithofer, BMW chairman, in a provocative statement. *Automotive News*, 7 March 2016.

3   Predicting Disruption

Gans (2017) questioned whether there are industries that are "ripe for disruption". He responds to this by first observing the existence of a paradox, in that if disruption events can be predicted, they are not really disruptive events. He adds that, "Prediction is useful if it can be turned into actions that matter" and concludes that there are two reasons why predicting a disruption does not make much sense. The first is that a disruptive event is driven by several sources and even if we can predict that it will affect a certain industry, we cannot predict how and when it will occur. The second difficulty is that a particular disruptive event could arise, but it will not necessarily weaken the competitive advantages on which a company's strategies are based.

In *Big Bang Disruption,* Downes and Nunes do not confront the problem of prediction, but identify three elements of the new technologies that drive disruption in various sectors of the economy: (1) the declining cost of creation; (2) the declining cost of information; and (3) the declining cost of experiment.

- Moore's law continues to exert its effects, reducing technology costs. Both the sourcing of parts and components and the delivery of products and services are now more reliable compared to the analog past. Even the costs of R&D are falling, "as idea generation, research and even innovation funding migrate to the cloud and to new form of incubators". Thanks to digital technologies, new companies can easily enter the market with products or services at lower costs and better customized compared to those of the incumbents (i.e., Uber, Lyft, Didi in the automotive industry, and Airbnb in the hospitality industry).[5]
- Thanks to the mass of information created "for consumer to consumer", it is now easier and more efficient to collect all types of information. New technologies enjoy exponential growth, thereby increasing the capacity of new companies to enter old and new markets, and clients of new entrants do not linger in the "early adopter" stage, as the classic marketing theory suggested.[6] Social networks and peer-to-peer information exchanges in fact immediately alert consumers to new products or services, and allow them to find out early users' opinions. "So there's little point to carefully timed marketing campaigns addressed to different customer groups over a control product release." This concept is expressed in other words in *Marketing 4.0* by Kotler et al. (2017) "Customers are becoming more horizontally oriented." They are ever less attracted by the appeal of the brand and "are relying instead on the f-factor (friends, families, fans and followers)".

---

[5]The text quotations are taken from Downes and Nunes (2014).

[6]Seeking to explain how, why, and at what rate new ideas and technology spread, Rogers (1962) developed a theory on the "Diffusion of Innovations" (also the title of his book). New products are not adopted as soon as they are launched. Some consumers are driven to buy new products as soon as they become available, while others prefer to wait before buying, so as to avoid wasting their money. Rogers identified six categories of adopters: innovators, early adopters, early majority, late majority, and laggards. Innovators are consumers that want to be the first to own the latest products. They are at the very beginning of the product life cycle. Early adopters are those that are open to new ideas, who prefer to wait until after the launch of the product. They make up a significant share in the early stages of the cycle.

- Global broadband networks and the diffusion of computing devices increases the possibility of collaboration among innovators and users. New products or services are introduced onto the market following simple, fast tests, carried out with real consumers, bearing costs and taking limited risks. Costly research is partly replaced with tests carried out directly on the markets with real users co-opted as collaborators. "When the right technologies and business model come together, the market takes off dramatically". The speed of diffusion of products and services that meet consumers' requirements is another distinguishing characteristic compared to the traditional theory of the "diffusion of innovations".

The possibility of experimenting and interacting with consumers, suppliers, dealers, and other actors in the automotive industry quickly and bearing low costs was also extended to the exploration of possible strategies, in particular marketing strategies.

Understanding potential threats is at the base of every strategy. The most common reaction among incumbents faced with the risk of disruption is to adopt a strategy of continuous change while waiting for the threat to manifest itself fully. In the automotive industry, however, the risk, which is evident in the case of digital technology, is that dynamic upstarts and digital giants such as Apple, Google, and Microsoft will gain dominant positions. Ford took the path of experimentation. Ford undertook over 25 different strategic experiments to try to understand what the future of transport might be when the industrial tradition of the past and digital technologies came together. In the words of the CEO of Ford: "We have given our engineers, scientists and technologists a challenge. We have asked them to use innovation not to just create better products. We have asked them to innovate to make the entire transportation experience easier to make people's lives better and, in doing so, to create a better world". In 2016, just a few months after the end of this series of experiments, Ford set up a new division: Ford Smart Mobility.[7]

Under the influence of digital transformation, the positions between incumbents and disruptors change quickly. In the automotive industry, new empowered entrants—Uber and Lyft—are now among the major incumbents, as Amazon and Netflix are in their industries. It is worth repeating the story of Apple's iTunes here, which successfully disrupted brick-and-mortar music retailers and was later disrupted by Spotify and its music-streaming business model. Few years have passed since the disruption Uber and Lyft had on the taxi industry, and now this industry could be disrupted again. "The advent of autonomous taxis and the wave of investment by automakers in car-sharing fleets could flip the economics of ride-hailing".[8]

---

[7] The textual quotation by the CEO of Ford is taken from Venkatraman (2017).

[8] "Will disruptors of the taxi market become disruptees?" Automotive News, 26 March 2017.

4   What Do Disruptors in the Automotive Industry Have in Common?

Simoudis (2017) has found that disruptors share "several important characteristics that he considers as critical to their success": (1) They are big data and machine intelligence companies, at their core; (2) they use direct-to-consumer models; (3) they are venture-backed startups; (4) they experiment constantly; (5) they make decisions with incomplete data and uncertainty; (6) they hire the best; (7) they are organised in small teams; and (8) they take risks and break rules (Table 1).

Table 1   Five companies at the core of automotive disruption

Tesla	Tesla is trying to disrupt the entire automotive value chain. Its innovations started with the electric connected vehicle, but also include several others. With the introduction of Autopilot, Model S and Model X are equipped with level 3 driver automation. Tesla is disrupting by utilizing big data. For the future, it has committed to offering level 4 driving automation. As its vehicles become increasingly autonomous, it is willing to offer a car-sharing mobility service and a ride-sharing service
ZipCar	ZipCar was the first to offer car-sharing mobility services. Its major inventions include a car-sharing business model that is connected to its big data software platform. Zipcar disrupted the car rental industry, which is the reason why Avis took over the company. Zipcar analyses the data it collects to identify new locations to place cars, to rebalance its fleets based on usage, to offer one-way rentals at more competitive prices than full-service companies, and to offer a lower price/hour of usage
Uber	Uber has innovated through its business model and with its mobile application, which combines software with big data with the broad utilization of big data such as in demand-based dynamic pricing and autonomous driving. After disrupting the taxi and limousine business, it is attempting to disrupt the entire on-demand delivery industry and to start competing with Amazon and Google. Through autonomous vehicles, Uber is attempting to control its costs in order to improve its margins and lower its prices
Google	Google is disrupting the automotive industries with two platforms. First, its Android Auto mobile platform that can control the car dashboard, including the navigation platform that is based on Google maps. In addition, Google has been developing a self-driving platform that can be used with level 4 and level 5 driving automation. Google is developing applications that offer dynamic ride pricing to optimize the usage of the network and the number of vehicles that are needed to serve a territory
Apple	Apple is disrupting the automotive industries with its CarPlay platform that controls the car dashboard, similar to Google's Android Auto. According to Simoudis (2017), "Apple can disrupt the automotive industry even if it chooses to focus exclusively on the user experience. It can disrupt not only the car's software and hardware platform but also the overall car-buying experience, car-servicing experience, etc. very much like it did with its mobile devices (iPod. iPhone, iPad)"

Analysis and synthesis of documentation from Simoudis (2017, pp. 63–69)

5 Identifying the Right Moment

The key problem for incumbents is when and how to react to a disruptive event. Choosing the most opportune moment to drastically change strategy is more critical than ever in an environment in which technologies are developing at an exponential rate.

The history of the automotive industry teaches us that on multiple occasions, some companies have predicted the arrival of a disruption and have invested considerable resources into tackling it, only to discover that it was not as dangerous as they had feared. Others understood what was about to happen only when it was too late (i.e., the financial crisis of 2008). The main reason is that in conditions of uncertainty, "it's far from clear to see and manage"… "The key to deal with disruption is to understand that it emerges surrounded by uncertainty" (Gans 2017).

As for when, this will likely be determined by two interdependent factors: when and how buyers will make their decisions, and when the manufacturers' proposal will have a product architecture or one of relatively stable services (dominant design).

It is necessary to begin with customer behaviour and customer expectations. As for EV, as is always the case, there are already those that want innovations, but their demand is obviously very far from achieving the manufacturer's break-even point. Only when the "dominant design" has been clearly defined and consumers can discover the product first-hand will it be possible to make reliable predictions about the future of demand.

How will they choose? Psychologists warn that in the short-medium term, it will not be a rational decision. Dan Ariely explains why in *Predictably Irrational*: "Humans rarely choose things in absolute terms", he explains. "We don't have an internal value meter that tells us how much things are worth". Instead, Ariel continues, we focus our attention on the relative advantage of one thing over another, "and estimate value accordingly". To decide, many of us not only need to compare one thing with another taken as a reference, but concentrate on things that can easily be compared, "avoiding comparing things that cannot be compared easily". The comparison made by buyers who are weighing up whether or not to buy will be between an EV and a traditional vehicle, but this will only be possible when the performance and prices of the new products have been clearly established and recognised. "Let me start with a fundamental observation", writes Ariely, "most people don't know want they want unless they see it in context. We don't know what kind of racing bike we want until we see a champ in the Tour de France ratcheting the gears on a particular model". This prediction of "predictable irrationality" applies to purchases by individuals and not to those purchasing fleets of autonomous driving vehicles and EVs, whose decisions are generally taken following a rational process.

6   How to Manage Disruption

Various authors have tackled this question. In particular, both Rogers (2016) and Gans (2017) have examined the reactions of incumbents when faced with the threat of disruption.

For Rogers, the first step is to predict the impact of a new disruptive business model on consumer behaviour. What might they adopt from the new offer? If the adoption were to spread, what would happen in the market? The second step involves assessing the likely scope of the offer and estimating how many consumers could move to the new product once it has been well-established. In the automotive industry, how many consumers would adopt a level 4 or level 5 driverless car (considering that they will be much more expensive than a traditional car)? The third step is to extend the analysis of the impact of a disruptive business model to other industries related to the one specifically of interest. Which other incumbents in other industries will be impacted, and how will they react? In the case of the automotive industry, the analysis needs to be extended to the supply chain and the channels of distribution.

According to Rogers, following the preliminary analyses, an incumbent has six main possible responses when faced with a disruptive business model: (a) three strategies to become a disrupter (geared at occupying the same spaces targeted by the disrupter); (b) three strategies to mitigate losses from the disrupter (an attempt to make the threat of the disrupter less hostile).

(a)   Three strategies to become a disrupter.

(1)   *"Buy out the disrupter"*. This is the simplest strategy. When the car rental company Avis understood that Zipcar's business model could disrupt the industry, it decided to buy it out. The rule is to keep the new acquisition distinct from the rest of the business. Zipcar is now independent from the rest of Avis. The new acquisition will continue to take away clients from the purchasing company, but the strategy could attract others.

It is not always possible to buy out the disrupter. The most frequent reason for this is the cost of the operation. If the disrupter has already consolidated its position, the price rises. For this reason, one frequently sees acquisitions of startups in their early days, including in the automotive industry. However, if the choice of target is not shrewd, there is clearly a greater risk of disappointment. An alternative for the incumbent is to purchase a shareholding in the capital of the disrupter, and not in its entirety. If the startup makes progress, the incumbent will subsequently seek to acquire total control. The risks go down, but the price goes up.

(2)   *"Launch a new disrupter"*. Rather than buying out the disrupter, the incumbent directly develops a new disruptive business model, seeking to beat the disrupter at their own game. This action must be taken before it is too late to take advantage of the new development of the industry driven by innovation. For the incumbent, this strategy will inevitably involve cannibalising their core business. The rules for fostering internal development in a startup apply here. They benefit from

being part of a group and using its resources, but they need to have independent strategies from those of the founder, a small and lean organisation, and remain isolated from the rest.

It is possible, but difficult, to launch a startup that is disruptive by anticipating the rise of other competitors. VW Group confronted this risk by challenging the ride-hailing company Uber with a new mobility services brand called Moia that will have its own electric passenger vehicle. Even Uber, on another front, followed this route, setting up a separate division to manage the self-driving technology (which is considered as a disruptor of the industry).

> **Dejà vu all over again.** In the history of the automotive industry, one can find examples of manufacturers that have sought to tackle the threat of new entrants by setting up a new organisational division that adopts the same technologies or the same marketing strategies as the new entrants. We might recall, for example, the experience of Saturn, founded by GM to defend itself against the Japanese competition of the 1980s. The results were modest. Saturn was later closed when it was overwhelmed by the consequences of the financial crisis of 2008–2012.

It is early to judge what will happen in the car industry, but in other sectors the results have been disappointing (for example, IBM, in the PC industry, set up a division that was then sold to Lenovo). Why did many attempts fail? It is partly because it is difficult to transmit knowledge and innovation from one division to the rest of the organisation, and partly because difficult conflicts arise between the management of the new division and those of the other divisions.

(3) *"Split the disrupter's business model"*. The third strategy involves finding other companies that detect the same threat of disruption, have the same objective to defend themselves, and partner up with them to re-create the disrupter business model.

   In the automotive industry, the major manufacturers have adopted this strategy. For example, VM invested in Ford-backed Argo. Both manufacturers understood that they were behind in the development of new technologies and that joining forces could cut costs and speed up recovery.[9]

   BMW and Daimler followed a similar path. Major rivals in the global luxury car sales sector decided to create a joint venture to build economies of scale in

---

[9]"Volkswagen and Ford have joined forces on electric and self-driving vehicles, capital-intensive areas that are reshaping the auto industry. The investment in Argo is significant, as it could accelerate VW and Ford's self-driving efforts, a costly but crucially important element of the auto industry's future. Both carmakers have been dogged by the perception they're lagging behind in developing the technology, and cooperation would allow them to share costs and potentially catch up faster". Source "VW may invest in Ford-backed Argo as alliance talks advance", *Automotive News*, 9 November 2018.

the new digital technologies and services, underscoring the transformation that is shaking up the automotive industry.

The main objective is to invest in the development of car-sharing and ride-hailing. The alliance also includes on-demand mobility parking services and electrical charging. Each company will own 50% of the joint venture. The head of auto research at UBS has estimated the value of the assets involved to be between €3bn and €9bn. He said that BMW likely paid Daimler to have a share in the joint venture equal to that of Daimler, since Car2go car-sharing (introduced by Daimler) is three times the size of BMW's DriveNow and ReachNow, brought together in the partnership. Some analysts have observed that in an age of shared mobility, in which the inclination for private car ownership has dropped and many people opt for the use of Uber, Lyft, and Didi, "carmakers could lose their customer facing brand value and become more like Boeing and Airbus".[10]

(b) Three strategies to mitigate losses from the disrupter.

(1) *"Refocus on your defensible customer"*. The incumbent reacts by focusing resources on those customers it has a better chance of retaining. It cannot continue with the same value proposition as in the past, which is now threatened by the disrupter. It therefore needs to refocus its offer to continue to create value for them. Marketing, particularly promotion, and innovation play an important role in this operation. Giving up on the clients most susceptible to the appeal of what the disrupter is offering means saving resources.

(2) *"Diversify your portfolio"*. If the chances of acting as a disrupter (stand-alone or in partnership) are slim, an alternative is to reposition the offer and potentially buy out small companies to speed up the entry into new segments or niches. A premise for success is leveraging on unique skills and assets. Rogers observes that diversification allows the company to leverage strengths as a whole, and although the profitability of the new areas of activity cannot be comparable to that of their core business, "they can create new opportunities for growth and can make the firm less susceptible to total disruption".

(3) *"Plan for a fast exit"*. When it is clear that it is not possible to oppose the threat of the disrupter through any of the above-mentioned strategies, the alternative is to quickly leave the industry (through transfer or liquidation). If the company is built up on multiple businesses, a spin-off of the business under attack and a quick transfer or liquidation could save resources for the entire company. The decisions are not easy, however. For example, if the business affected by the spin-off contributes to economies of scale, it is clear that its abandonment will lead to diseconomies of scale. Moreover, the abandonment of core technology can mean handing over a market share to traditional competitors, who instead seek to withstand disruption.

Gans (2017) observed that decisions like the ones summarised thus far are difficult to take and can take time for two reasons: uncertainty and cost of reaction. The

---

[10]"BMW and Daimler join forces to bolster tech prowess". *Financial Times*, 29 March 2018.

uncertainty derives from the fact that management can detect the threat of a new product or service, but this threat will not necessarily prove significant. The offer of the disrupter can attract demand, but not to such an extent that it severely threatens the incumbent. The problem is that management can arrive at this appraisal only after a certain time, during which uncertainty presides. The second observation is that whatever the response to the threat is, it always involves a cost, and this too encourages waiting before reacting to the threat.

Ghosn, CEO of Renault-Nissan, attributed the fears of a disruption of the automotive industry to the speed with which the digital transformation erupted. He declared that he was not afraid of Apple or Google, even though they have significant resources to invest and are now developing autonomous-drive vehicle technology. He was sure they would not enter the scene as manufacturers. "Those companies are now wealthy enough to buy automakers outright if they choose to become carmakers". "They could have done it already".[11]

References

Christensen C (1997) The innovator's dilemma. Harvard Business School Press
Downes L, Nunes P (2014) Big bang disruption. Portfolio, Penguin
Gans J (2017) The disruption dilemma. The MIT Press
Kotler P, Kartajaya H, Setiawan I (2017) Marketing 4.0. Moving from traditional to digital. Wiley
Rogers E (1962) Diffusion of innovation. MacMillan
Rogers DL (2016) The digital transformation playbook. Columbia Business School
Simoudis E (2017) The big data opportunity in our driverless future. Paul Lienert
Thompson B (2013) What Clayton Christensen got wrong. Stratechery, September 22
Venkatraman V (2017) The digital matrix. New rules for business transformation through technology. LifeThree Media

---

[11] Apple and other super-rich tech companies should not be feared as auto industry disrupters, Carlos Ghosn says. They should be embraced. "There's been a lot of talk about disruption, about new competitors who promise a new approach to what a car can be," Ghosn told an industry audience. "Much of this unease over potential disruption has resulted from the rapid emergence of new technologies and so-called mobility services. Ghosn argued that tech giants such as Apple, which is now developing autonomous-drive vehicle technology, do not want to be automakers because the profit potential is too small. "Don't fear the tech giants", Ghosn says. *Automotive News*, 23 March 2016.

Chapter 16
How Platforms are Reshaping Automotive Marketing Management

Abstract On the threshold of the new millennium, certain carmakers made an important initial reaction to the advances in the digital economy. They started to prepare themselves for their future as "transportation solutions providers", and no longer just vehicle manufacturers. The "day of reckoning" came about in November 1999 when both Ford and GM announced their plan to each launch their own value chain based on digital technology. However, a new breaking point in the marketing of automotive companies was approaching. It was driven by new digital technologies that enormously expand the reach, speed, convenience, and efficiency of platforms. It derived from the fact that the system of relationships rendered possible by the platforms changed and damaged the traditional value chain in the automotive industry. The new technologies facilitated communication and the exchange of data between participants in the network. The more a company attracted new participants to the platform (owners, providers, producers and consumers), the greater the network became and the more transactions between demand and supply increased. The larger the scale, the more value it generates. The capacity of certain companies to take advantage of the development of the platforms soon posed a threat for incumbents. New competitors quickly entered the market, proposing a new kind of "crowd-based public-private partnership", including Uber, Lyft and BlaBlaCar. The emergence of Tesla is also a clear example of a threat for incumbents. In recent years, digital marketing in the automotive industry has made significant advances, but it is still behind compared to other major sectors.

On the threshold of the new millennium, several carmakers, including Ford and GM, made an initial important reaction to the advances in the digital economy. They started preparing themselves for their near future as "transportation solutions providers", and no longer merely vehicle manufacturers. The proceedings from a conference held at Wharton University in 2000 helped understand what was happening. It was a widely held view among the speakers that modern organisations in the digital age needed to be constructed on new principles and business models and that marketing was among the functions most affected by the need to change. The proposal put forward by the companies' management was standard practice for those confronted

E. Candelo, *Marketing Innovations in the Automotive Industry*, International Series in Advanced Management Studies, https://doi.org/10.1007/978-3-030-15999-3_16

with a high degree of uncertainty: understanding what was changing and acquiring the capacity to respond with flexibility, agility and speed.

In a few years, advances in technologies became explosive and the digital age spread across every part of the scene, even in the car industry, especially in marketing. The use of digital platforms in the automotive industry and the emergence of network effects forced marketing management to face new challenges. One consequence of the network effects was the deconstruction of the value chain, that is, the model that had led carmakers' strategies for decades. In the digital age, advances in technology created advantages not primarily on the side of the supply (supply-side economy of scale) as in the past, but rather on the side of demand, and generated significant demand economies of scale.

Compared to other industries, carmakers were slow to respond to the effects of digital economies. Anyone who was slow to react and then sought to make up for even a small delay had to make heavy investments in a stage in which great uncertainty about the future made it difficult to establish secure objectives and paths.

1   The Day of Reckoning

After twenty years of deferments, some carmakers finally decided to act. The "day of reckoning" occurred on 2 November 1999, when Ford and GM each announced their plan to launch their own value chain based on digital technology.

AutoXchange. It was a stand-alone joint venture between Ford and Oracle, which offered supply-chain services to everybody, including Ford's competitors. For its part, GM introduced CommerceOne, which offered a bundle of services called market-size. Unlike Ford, GM limited its market-size offer to its own suppliers. Another step forward, even more important from the marketing perspective, came about a few months later during the 2000 Detroit Auto Show. Both Ford and GM announced their "in-vehicle internet services". Ford announced a joint venture with Yahoo and GM with AOL. Ford's new customer proposition was particularly well-structured. Nasser, CEO of Ford, declared that the objective was to transform the company into a "transportation solutions provider". Ford had recently purchased Kwik-fit, a British chain of two thousand service centres all over Europe. Above all, Ford made an agreement with Priceline and with MicrosoftCarPoint, granting these two web intermediaries the role of first point of contact for potential car buyers, thereby marking an innovation in marketing (Tapscott et al. 2000).

2 A New Way to Think About Marketing

> Today, a new technology is challenging the fundamental basis of traditional marketing discipline. The Internet is transforming not only the practice of marketing but the way we think about marketing. It is turning marketing on its head. (Wind and Mahajan 2001)

These were the words of Wind and Mahajan in the first lines of *Digital Marketing*, a book that gathers together the papers presented at a conference held at Wharton University in 2000. It drew together experts from the most prestigious universities, such as Wharton itself, as well as Columbia, Harvard, Northwestern, and MIT.

The challenge for the change triggered by new technologies came from afar. It did not happen overnight, as we suggested in Chap. 14. However, at the beginning of the new millennium, it already seemed deep and irreversible. Following a classic model in academic research, the conference at Wharton was divided into three parts: (1) the impact of new technologies (e-commerce in particular) on marketing and marketing research; (2) how the digital environment was changing consumer behaviour and demanding new approaches to marketing research; (3) how companies could meet the new challenges by implementing new marketing strategies focused on communication, customization and pricing.

Even in the car industry, the change had been underway for some time, as illustrated by the birth and success of Autobytel.com (Chap. 14).

The speakers at the conference asked: what were the consequences? What came to be defined as the digital revolution, giving access to a global audience of users and further shifting the power from producers to buyers, broke down barriers between sectors and led to the emergence of a "global grid"—"a network of users and portals"—subsequently called a platform.

While in a stable or relatively stable environment planning is the base for building long-term strategies, under the impetus of advances in information technology the environment changed so quickly that many solutions soon became obsolete. Wind and Mahajan (2001) have compared the characteristics of the new business models considered as emerging with the traditional ones that had dominated competition in preceding decades. They have identified the following transitions from the old to the new model as follows: (1) from technology as an enabler to technology as a driver; (2) form seller-centric models to customer-centric models; (3) from physical to knowledge assets; (4) from vertical to virtual integration; (5) from functional-centric to cross-functional integration; (6) from planning processes to experimentation and learning; (7) from decreasing returns to scale to increasing return to scale.

The model of the corporation, which Alfred Sloan had made famous and which had been the "engine of progress for decades", was now considered as "a burden in the age of the virtual corporation." Modern organizations in the digital age need to be constructed on new principles. Faced with increasing uncertainty, companies need to respond with agility and flexibility. And to create agility and flexibility, new business models are required.

The new business models need to be flexible enough to encompass relationships between the company, its suppliers and its customers, relationships which change on a frequent basis and in a way that is difficult to predict. They must also react quickly to changing situations. After just a few years, history confirmed this need with the rapid decline in the Dell model (which had very successfully developed built-to-order models in the personal computer market) and that of Kodak (although it had first developed digital technology in photography).

3   Digital Technology Pervades Management Practices

In the previous pages (particularly in Chap. 14), we identified the first effects of the transformation initiated by new technologies since the mid-1990s. In the early 2000s, the digital age also spread to the car industry, and its effects played a key role in the agenda of anyone wishing to adapt marketing management to advances in technologies.

However, some people would ask: digital computers had been in use since the 1960s, so why are we only now talking about the digital age, so belatedly? McAfee and Brynjolfsson's response in *Machine, Platform and Crowd* (2017) was that the digital age was discussed from the point at which digital technologies started to be frequently used in management, from production to marketing, finance to R&D, and marketing research to customer behaviour analysis.

The two authors identified a first and second stage in the acceleration of the digital age. The first stage began towards the mid-1990s; and when digital technologies began to have a significant impact in two fields: companies and the economy. In companies, it occurred by largely replacing repetitive activities like "processing payroll, welding car body parts together, and sending invoices to customers". In the economy, it occurred when computers and other digital technologies emerged as principal factors of the growth in productivity.

The beginning of the second stage, according to the two authors, is more difficult to identify, but the signs of its emergence are to be found in "science fiction technologies, in the stuff of movies", in digital editions of books and magazines. In the automotive industry, it occurred later when, in 2010, Google announced that completely autonomous cars were in circulation in California. It occurred when, on the TV, an IBM supercomputer beat two human champions in a quiz programme.[1] And it occurred again when, in the early 2010s, smart phone sales exploded.

---

[1]In 1997, IBM's Big Blue computer beat the world chess champion at the end of a competition spread across six meetings: two won by IBM; one by the champion; and three draws. Driven by this success, IBM researchers designed a machine able to beat the champions of an even more complicated game known as *Jeopardy!* In 2011, an IBM computer called Watson (from the surname of a CEO of IBM of the "roaring twenties") beat two of the most successful players of all time.

4 The Rise of the Platform

A new breaking point in the marketing of automotive companies shook up the structure of the competition and marketing strategies. It arose from the advances in technologies that spread digital platforms into the automotive industry, and from the consequent emergence of network effects.[2] The consequences in the automotive industry were considerable: "As a result of the rise of the platform, almost all of the traditional business management practices… are in a state of upheaval. We are in a disequilibrium time that affects every company and business leader" (Parker et al. 2016).

Platforms had existed for years. They connected magazines with their subscribers, for example, and with those purchasing advertising space. Even Autobytel.com used a platform (see Chap. 14). In the lead-up to the 2000s, new technologies made it easier and cheaper to scale up platforms, and considerably increased the capacity to collect, analyse and exchange data. The platform connected producers and consumers in an interaction that generated value for both parties.

Others had used the expression ecosystem to indicate the content of what is now called a platform. This included James Moore who, in *The Death of Competition*, which came out in 1996 (new edition 2005), introduced "biological ecology as a metaphor of strategic thinking about business co-evolution and radically new co-operative/competitive relationships."

As noted by Marshall et al. (2016), the platform connects four types of players with different, complementary roles. "The *owners* of platforms control their intellectual property and governance; *providers* serve as the platform's interface with users; *producers* create their offerings, and *consumers* use those offerings."

5 Deconstructing the Value Chain

Urged on by new technologies which "enormously expand a platform's reach, speed, convenience, and efficiency", a turning point came about when the system of relationships rendered possible by the platform changed/undermined the traditional value chain in several sectors. This initially occurred in a restrained manner in the automotive industry, in a more forceful way in many others, and in a devastating manner in others (e.g. entertainment and publishing). "Internet and associated technologies give today's platform businesses a truly breathtaking ability to transform industries, often in unpredictable way" (Parker et al. 2016). For decades, the dominant model of industrial transformation, and of the car industry in particular, was based on the sequence of interlinked activities which led to the creation of a higher value than that of the production factors used. The more appropriate the strategic choices were, the more value was created.

In the traditional value chain, value is created in each of the various stages it comprises. In the automotive industry, for example, the traditional value chain comprises:

---

[2]"*A platform can be described as a digital environment characterized by near-zero marginal costs of access, reproduction, and distribution*" (McAfee and Brynjolfsson 2017).

(1) parts and component suppliers (which create value with higher value products); (2) automotive manufacturers (which create value through assembly, design, marketing and branding); (3) automotive dealers (which create value through sales and inventory management, services pre- and post-purchasing); and (4) consumers.

When a platform consolidates its position and renders the exchange of relationships between its participants more intense, it deconstructs the traditional value chain. In *Digital Capital* (2000), Tapscott et al. were among the first to describe the reasons for and modes of this passage. "A new logic of value creation now drives the production of goods and services. Demand-driven value chains are replacing old make and sell, push models, as producers 'build b-web' that link and respond directly to consumers." They wrote that value chains transform raw materials and other production factors—"atoms in the physical world and bits in the electronic world"—into finished goods and services. In the chain, value is added in each of the stages of activity that span from the availability of production factors to making the product or service available to the client. Tapscott et al. complete their analysis by distinguishing two types of value chain, which can be used in a combined manner: (1) routine production; and (2) shop production.

Routine production. Product-centric. The objective is to make, move and market physical goods. "The make-and-sell logic of physical capital drove the old production model." Management seeks to predict trends in demand and to consequently plan production rates and volumes. In the short term, managing stockpiles became the primary tool of flexibility. In this type of production, carmakers have considerable control over the value chain. For example, Ford and GM used automated linkages to their closest suppliers to transfer their inventory management issues to them.

Shop production. Typical of professional services, construction design, research and development. The company responds to a particular demand, which could be unique (i.e. management consulting firms). Shop production has three characteristics that distinguish it from routine production:

(1) It does not fully repeat what might already have been done and does not propose standard procedures. Value-creating activities are unique.
(2) The cycle is entirely triggered by demand. It is the client that takes the initiative. The product or service is produced only after it has been paid for or the client has committed to paying, with no inventory.
(3) The client participates in the design and configuration of the product, and sometimes partakes in its creation (as in a physiotherapy rehabilitation project, for instance).

The reference was Dell's build-to-order business model, which was highly successful in those years and which Tascott et al. considered as a hybrid between the two forms of production. Clients ordered their personal computer directly the producer (Dell), sidestepping the distribution channel. Without an order, Dell would not produce. The client configured the product by choosing from a limited number of elements to be combined (20 elements in Dell's case).

6 The Power of Network Effects

To explain how the traditional value chain is deconstructed and with what consequences, Marshall et al. (2016) compare the new economy with the industrial economy. The driving force behind the industrial economy was (and is) the supply-side economies of scale. The use of technologies (mainly those of the assembly line) that allowed for mass production gave Henry Ford the drive to enjoy a rapid, notable success. The larger the production volumes, the lower the product costs per unit and the marketing and distribution costs. For Henry Ford, the advantages of controlling the supply-side economy of scale lasted until customers accepted a standardised product and were attracted by the low price. Failing to understand the change in customer behaviour signalled Ford's inexorable decline. On a different scale and with different people involved, this error has been repeated various other times in the automotive industry.

In the digital age, advantages are generated by advances in technology that have an effect not primarily on the supply side (supply-side economy of scale), but rather on the demand side, generating significant demand economies of scale (as already explained). The new technologies facilitate communication and the exchange of data between network participants. The more a company attracts new participants in the platform (owners, providers, producers and consumers), the greater the network becomes, along with the transactions between demand and supply, and "the richer the data that can be used to find matches". The larger the scale, the more value it generates because the average value of transactions between participants in the platform increases. An ever-greater scale attracts new participants, who create new value to the point of establishing positions of near-monopoly in the most successful cases. These network effects have in fact granted dominant positions to Alibaba, Google and Facebook.

To sum up the innovation briefly, the companies that adopted the traditional value chain model (described by some as the pipeline model) focused on increasing sales, while those that adopted the platform-based model focused on developing relationships between participants in the network.

Recent history demonstrates the destructive power of network effects. "When a platform enters the market of a pure pipeline business, the platform virtually always wins" (Marshall et al. 2016).[3]

---

[3]Marshall et al. recall that: "Back in 2007 the five major mobile-phone manufacturers—Nokia, Samsung, Motorola, Sony Ericsson and LG—collectively controlled 90% of the industry global profits. That year, Apple's iPhone burst onto the scene and began gobbling up market share". By 2015, the iPhone had singlehandedly generated 92% of global profits, while all but one of the former incumbent made no profit at all.

7   The Road Ahead in the Car Industry Looks Bumpy

The capacity of some companies to take advantage of the development of the platforms quickly proved a threat for incumbents in the automotive industry. The emergence of Tesla is a clear example of this. It removed parts of the traditional value chain and redesigned other parts of it (production and distribution, in particular).

> **Tesla.** "The deconstruction of the value chain is possible because Tesla has chosen to build intelligent machines instead of just automobiles. The data generated at each step enable the company to optimize sales, support, and services, thereby making it economical for a single company to be vertically integrated along the value chain [...] With the intelligence provided by Tesla cars, there is little need for automotive dealers. Hence, Tesla has built a business model that excludes traditional dealer networks, instead opting to build their own company-owned dealerships" (Thomas and McSharry 2015).

The construction of Gigafactories by Tesla to produce batteries for its vehicles, also from the perspective of a latent market, was a further step towards the deconstruction of the traditional value chain of the automotive industry. It deconstructed in particular the supply side of the value chain. Later we knew that towards the end of the 2010s, the Tesla's business model would have raised many doubts in terms of reliability and profitability.

8   Reasons for a Late Response

Why were carmakers slow to respond to the effects of digital technologies? For four reasons: (1) The effects of innovation in the industries of durable goods came later; (2) Most manufacturers did not understand what was happening and underestimated the possible consequences, e.g. GM especially; (3) Manufacturers were afraid that an overly fast change might damage existing products; (4) The development of digital technologies was explosive.

- Companies in the durable product sectors, capital intensive, like those of the automotive industry, experienced the effects of digital technologies later than those of other sectors, for example compared to those subject to fashion trends. The lifecycle of a durable product is longer than the average for non-durable products, and consequently the period in which the buyer maintains the use and ownership of the product is also generally longer. The demand for durable goods, such as the automobile, is therefore less vulnerable to rapid changes in "demographics and social attitudes" and technologies.

- In recent history, too, the management of certain automotive firms has proven to have failed to understand that the threat was looming and that it would soon manifest itself. The financial crisis of 2007–2008 had sped up the demand for fuel-efficient cars and cars that, with lower harmful emissions, guaranteed better environmental protection. European car companies (particularly the premium brands), Japanese companies (particularly Nissan), and the American company Ford had interpreted the warning signals of a new trend. The environment variables had become more volatile, and competition, in particular, had become more intense. The crises of Dell, Nokia, Kodak and other icons as a result of their having lost contact with the evolution in technology was a warning to the marketing management of the car industry; not everyone picked up on it, however, including GM.

> **General Motors.** For 70 years, it had sold more cars than any competitor. It had survived the Great Depression of the 1930s, the crisis following World War II and the drastic increase in the price of gasoline in the early 1970s. In 2009, it was saved by the Obama administration and forced to undergo a dramatic downsizing. Historic brands such as Oldsmobile and Pontiac were abandoned, as well as Saturn which, twenty years beforehand, had been introduced to beat Japanese competition in the American market, and the Hummer truck, once a status symbol of the rich and famous, which had quickly become a symbol of waste and excess. The fact that management had not understood how customers' expectations had changed was clearly demonstrated during the hearings before the U.S. Senate, when the CEO of GM, Wagoner, repeatedly attributed his company's serious losses to external events such as the 2008 financial crisis, proving his ignorance of the fact that, for some time, GM had been leading 280 thousand employees into the abyss of bankruptcy as a result of products that no longer met customers' expectations.

- As in other industries, carmakers struggled to transform the business model and reinvent themselves because they were afraid of damaging or weakening existing products. On the contrary, the new competitors driven by the new technologies had no obstacles, or fewer obstacles. In this case, too, Tesla is a good example. It quickly entered the scene, achieving a higher market capitalisation than other American manufacturers combined. One of the advantages of Tesla was that as a new entrant, from the onset it introduced the latest progress in digitisation free from previous obstacles that weighed on incumbents.
- Most carmakers were very behind in embracing new technologies. As already noted, to make up for even a small delay, the exponential development of digital technology required car companies to make significant investments.

References

Marshall W, VanAlstine M, Parker G, Choudary S (2016) Pipelines, platforms, and the new rules of strategy. Harvard Bus Rev

McAfee A, Brynjolfsson E (2017) Machine, platform, crowd: harnessing our digital future. WW Norton & Company

Moore J (2005) The death of competition. Harvard Business School Press

Parker G, Van Alstyne M, Choudary S (2016) Platform revolution. Norton & Company

Tapscott D, Ticoll D, Lowy A (2000) Digital capital: harnessing the power of business web. Harvard Business School Press

Thomas R, McSharry P (2015) Big data revolution: what farmers, doctors and insurance agents teach us about discovering big data patterns. Wiley

Wind J, Mahajan V (2001) Digital marketing. Wiley, London

Chapter 17
Digital Technologies are Rewriting the Old Rules of Marketing

Abstract Advances in new computing technologies to gather and process large quantities of information, increased further in the second part of the 2000s. Nowadays, car clients make great use of digital media. Given that potential clients find a huge amount of content online, they make many decisions concerning the purchasing process before visiting a dealership. The marketing content they can find online has a significant impact on these decisions. The development of digital technologies marked the need to extend the "4Ps" of marketing to 7, to include people, process and physical evidence. The impact on the traditional marketing was profound. New technologies made it possible to change the core product and extend it. From the early 2000s, car companies moved towards "selling personal transportation solutions" rather than "just selling cars." Customers participated in creating products through the web. In the pricing field, transparency increased, downward pressure on prices grew, and dynamic pricing challenged vendor-fixed pricing, especially for the low-cost segments. Regarding place, the relationship between customers and dealers is difficult to replace. Almost everywhere, franchisees have obtained state legislation to protect their positions. However, given that many car buyers make their purchasing decisions before they even consider visiting a dealership, manufacturers often turn to augmented reality (AR) and virtual reality (VR) to present their products. In terms of promotions, the digital age opened up new channels of communication: some people speak of "the end of an era." In the digital age, customers have access to a large amount of information about products, and power shifts towards them.

1 Marketing in the Time of Automakers' Vulnerability

We have repeatedly remarked that in industrialised countries, the demand for automobiles is declining or at least not rising, while in emerging countries (BRIC, MINT) it is on the rise, driven by economic development and the consequent demand for mobility. Let us now take a detailed look at the causes for the stagnation/fall in demand in industrialized countries, and the effects on car marketing.[1]

---

[1] These initial pages follow the trace of Sumantran et al. (2017) and Oge (2013), and draw on data from Automotive News.

© Springer Nature Switzerland AG 2019   141
E. Candelo, *Marketing Innovations in the Automotive Industry*, International Series in Advanced Management Studies, https://doi.org/10.1007/978-3-030-15999-3_17

- The cost of owning a car is increasing. Over the past 50 years, the cost of owning a car has increased more than people's average income. In the U.S., in 1950, to purchase an average car, a family would have spent 45% of their annual income. In 2014, to purchase a car, the equivalent family would have spent 61% of their income. As well as the declining rate of growth of the global economy compared to previous decades, other factors have negatively impacted car demand: increased tax on car ownership and the cost of using it. On the eve of the 2020s, depressed incomes and climbing costs of car ownership and use are putting individuals off owning a car.
- Many young people prefer to have an internet connection. Having grown up with social media, many young people are used to freely expressing their personality and to having easy access to products and services. In a survey carried out in the United States, two-thirds of college students said that they would prefer to have internet access than the use of a car. For them, the car has become commonplace, no longer a novelty. It is no longer their first possession objective as it was for their parents. This is clearly true for those who have a sufficient family or personal income to be able to purchase a car. It does not apply to those who would like to buy a car but cannot afford to do so.
- Millenials and their aspirations: Millenials make up around a quarter of the population in the U.S. and in Europe. The attitude of these "digital natives" towards mobility is shaped by their preference for city living, their familiarity with technology, and the uncertainty they associate with their own economic futures. According to a Nielsen survey, "Millenials are less interested in the car culture that defined Baby Boomers." In the U.S., Europe, and Japan, the number of young people that take their driving licence or have access to a car is in decline. The number of Germans aged 25 and under getting a driving license dropped by 28% in the past decade. Consequently, fewer drive.
- Extension of public transport. Research found that in Europe, 23% of all journeys are made by public transport (bus, tram, metro), which is considerably higher than the percentage in the U.S. This is largely due to the efficiency and extensiveness of the networks, but the tendency across Europe is towards an increase in the use of public transport. In Europe, the average journey times to get to work are shorter than in the U.S. because the cities are more compact. Public transport allows commuters to work or entertain themselves with their smartphones or social networks while travelling. Car use is thus reduced, as are the overall distances covered by commuters.
- The more you drive, the more problems you have. Driving takes an increasing amount of time for commuters. Traffic and the difficulty of finding a parking space at their destination have a negative impact. According to Sumantran et al. (2017), there seems to be a psychological threshold for tolerating distance and time when commuting by car, above which individuals have to find alternative solutions.
- Changing population demographics. The need for mobility diminishes with age. In industrialised countries, older people make up a significant proportion of the population, and this proportion increases more quickly than for young people. This

trend also reduces their aspiration to own a car. This is not the case in emerging countries, where young people make up the largest proportion of the population, and where the working population is on the rise.

- Growth of e-commerce. One of the factors that encourages people to use a car is shopping. The diffusion of e-commerce is therefore another factor that reduces car usage. The strength of Amazon, Alibaba and other such companies is demonstrated by the difficult position in which they have placed retailing giants such as Walmart. Even the tendency to eat meals in one's own home reduces car use. The same considerations apply to going to the cinema. Attendance has diminished as a result of the growing, varied offer of TV and streaming movies.
- The sharing economy. For "digital immigrants", sharing property is often a problem; this is not the case for "digital natives." This explains the spread of car-sharing. BMW's own estimates show that over the next decade, one car-sharing vehicle will replace at least three privately-owned ones, and mobility services, including autonomous cars, will account for a third of all trips. This trend is advantageous for protecting the environment because it reduces the number of cars on the road, and from the economic perspective, inasmuch as it allows for a greater use of assets; however, this too reduces car use.
- In *Driving the Future,* Oge identified four trends that "will determine what and how we will drive over the next few decades." They are: the growth of urbanization and megacities; the potential arrival of demand-side peak oil; the international convergence and adoption of greenhouse gas regulation; and the social impact of connected living and working.

Structural decline. The above-mentioned trends are accelerating the shift toward what is being dubbed the "peak car", "a time in the not-too-distant future when sales of private vehicles across the West will plateau before making a swift descent".[2] This is true, in particular, for major cities in which people are inclined to share means of transport with others (ride-hailing, car-sharing, and other forms), rather than owning their own car "that sits idle most of the time".

The falling demand for cars (in industrialized countries) is taken for granted by many (Canzler and Knie 2016). Berylls Strategy Advisors predicts that by 2030, in the U.S., where data are most readily available, the "total sales of cars—individually owned and shared—will fall almost 12% to 15.1 million vehicles". The main problem is that if this happens, it will involve a structural decline never seen before, when contractions in demand were due to temporary factors such as an economic crisis. The responses provided by manufacturers to date are far removed from the breaking points. In other words, defending against the threat is already slimming profits and thus also the possibility of making investments.[3]

---

[2]Textual quotations taken from "Why the industry must prepare for the 'peak car' era", *Automotive news,* August 17, 2018.

[3]"Take the DriveNow car-sharing service BMW started in 2011, which charges users by the minute to rent more than 6000 BMWs and Minis in 13 European cities. After seven years, it's still turning a loss, and last year (2017) made up just 0.07% of the company's sales. The rest came mostly from

Let us examine the effects of the digital transformation on two marketing elements—marketing research and Ps—and management responses.

2   A Change of Tune in Marketing Research

The growing deployment and application of digital technologies has also brought about a profound transformation in marketing research. The most obvious result of this is the explosion of data regarding customers' buying behavior. In traditional marketing research, it is common practice to distinguish between two types of research data collection: primary data and secondary data. Primary data are collected in the context of a specific project using various methods (online surveys, online focus groups, observation methods, lab and field experiments, and others), while secondary data already exist in the external environment and must be identified and selected. Digital technologies have significantly increased the capacity both to collect primary data and to access secondary data. From the early 2000s, Burke et al. (2001), as well as identifying the many ways in which digital technologies have transformed marketing research, observed a rising scepticism about the role and efficiency of traditional marketing research in managerial decision-making, especially in relation to quantitative research. Many managers complained that marketing research often did not provide the information expected, cost too much for the information it offered, and when it did provide useful information, this often came late. According to the authors, "The digital medium offers an opportunity to address this limitations of existing research approaches." Advances in marketing research over subsequent years proved them right.

The amount of data at automotive manufacturers' disposal has also risen, well beyond the capacities of the human mind to manage it. They are thus faced with the problem of quickly selecting the information that is really useful. It is critical to understand who their audience is and how its members behave in the buying process. Given that potential clients can find a huge amount of content online, from the product characteristics to comparisons between products and prices, they make many decisions in the buying process before even visiting a dealership. These decisions are strongly influenced by the marketing content they may have found online. Moreover, it is not enough for car manufacturers to concentrate on current targets for customers reached; they need to understand those who are consuming information and dealing with brands digitally.

Research into what drives people to own a car, and what type of car, became much more difficult with the advent of the digital transformation, which has given buyers more power. The literature on this matter is very rich and offers ideas that are not shared by everyone but which are undoubtedly of interest.

---

selling almost 2.5 million luxury vehicles, such as the BMW 3-series sedan." Source: Why the industry must prepare for the 'peak car' era, *Automotive news*, 17 August 2018.

"Conspicuous consumption." Revisiting veblen. The relationship between social classes and the desire and possibility of owning a car, to which Alfred Sloan had wisely responded with winning marketing strategies, had long waned, but recently the importance of examining consumer behaviour in social classes has arisen again, with concepts rooted far back in time.

In *The Theory of the Leisure Class,* written in the late 1880s, Thorstein Veblen (1899, new ed 1924) defined his view of the relationship between material goods and status. He is famous for his concept of "conspicuous consumption," the use of particular goods through which the "leisure class" revealed its status. Veblen has harshly critiqued the upper classes of society of his time, accusing them of almost exclusively responding to social rank.

In *The Sum of Small Things. A Theory of the Aspirational Class*, Elizabeth Currid-Halkett (2017)—after observing that the leisure class no longer exists, having been replaced by the new dominant cultural elite (that could be called the aspirational class)—argues that, "In the 100-plus years since Veblen's book was published, his theories apply more today than ever before, and they apply to all of us." Currid-Halkett maintains that "conspicuous consumption" is still an important means of revealing one's social position, but that the members of the aspirational class "have found new means to show their status." It is not the super-rich that rely on cars to demonstrate their status, but rather the middle class, who "as a share of their expenditures is spending more on conspicuous consumption relative to their income while the wealthy (and the very poor) are spending less." Among the rich, conspicuous consumption has been replaced with spending on "non-visible, highly expensive goods and services." Partly similar concepts have been expressed by other scholars. For example, in *The Affluent Society* John Kennedy Galbraith observed that when many people can afford luxury goods, these goods lose some of their mark of distinction, and Vance Packard, in *The Status Seeker*, argued that material goods do not reveal status.

3 A Break with Tradition in the Marketing Mix: From Four to Seven Ps

The development of digital technologies affects all elements of the marketing mix, and raises more varied, rapidly evolving challenges. It also signals the need to extend the "P"s of marketing to seven. This would include not only product, place, price, and promotion (4Ps), but also people, process, and physical evidence. In essence, it is necessary to move from a traditional marketing mix (which primarily concerns tangible products) to a services one (with a prevalence of intangible elements) because the car is ever less a mechanical device and increasingly a means of providing services.

Product. Digital technologies make it possible to change the core product or extend it. The core product is the main benefit that buyers expect from a purchase, by means of which they can satisfy their expectations and needs. In the car industry, this is transportation from points A to B. An extended or augmented product refers to the benefits and services that are built around the core product. In the car industry, these are style, safety, status, fuel economy, and others. Digital technologies have further expanded the possibility of extension. From the early 2000s, car companies started to redefine their value propositions. They moved towards "selling personal transportation solutions" rather than "just selling cars."

The other main variations that the internet allows are: options to offer digital products (i.e., GM's OnStar system links automobiles with service centres to provide roadside service assistance), conducting research online, speeding up new product development, and speeding up the diffusion of new products. Customers now participate in the creation of the product. Through the web, clients can contribute to the creation of the product (built-to-order cars). For this reason, more so than in the past, cars in the digital age encompass the experience, knowledge, and tastes of customers.

With the digital age, some rules have taken on particular importance. The premise for success is to offer a product that must meet a real need in the market. More so than in the past, it is important to actually give the client what was offered with the value proposition. The client must find what they were expecting. Not offering this could result in a backlash as a result of the viral nature of the internet, which can quickly and easily "bite you." Manufacturers must be as sure as possible that the product and the entire chain related to it (pre-post-sale, accessories, complementary services) meet the clients' expectations as closely as possible.

The concept of "whole consumer experience" has also become important. This includes the research that precedes the purchase and valuation, the purchase transaction, and the use of complementary products and services. According to Wind and Mahajan, "Product designers can no longer assume that their responsibility ends with the purchase transaction."

For all these reasons, and others that we will see in the next chapter, the traditional product definition and product marketing model has partly changed.

Price. In a single number, the price sums up the client's estimation of an entire set of conditions: time of availability of the product, innovation, fashion, status, rarity, and long-term value, as Tapscott observes in *Digital Capital*. The effects of the internet's capacity to influence the circulation of information regarding pricing became quickly apparent. JD Power and Associated (2000) revealed that, in 1999 in the U.S., over half of car purchasers had consulted the internet during the purchase process.

Digital technologies have strong implications for pricing policies. According to Chaffey and Chadwick (2016), the main implications are:

- Increased price transparency, as a result of the greater amount of information at the customer's disposal. The internet equals transparency. In the digital age, it is easy to make comparisons. Ryan (2015) notes that, other things being equal, price is not sufficient to compete, even the lowest price. It is necessary to be sure that the entire value proposition is attractive for the customer.

- Downward pressure on price, deriving from the increased price transparency and the growing number of competitors. Downward pressure has also driven many products and services towards commoditization.
- New pricing approaches (including auctions). Many of these options were also used before, but digital technologies have made some models easy to apply. In particular, traditional or forward auctions (B2C) and reverse auctions (B2B) have become more widely used than before. In the automotive industries, the effects have been felt in negotiations with fleet companies.
- Alternative pricing structures or policies. The internet has made different types of pricing possible, particularly for digital downloadable products (i.e., new options such as payment per use, rental at a fixed cost per month).

Harsh lessons from the digital age. Simon and Schumann (2001) have identified five management lessons in relation to pricing policies in the digital age.

1. "The fundamental driver behind commerce of any kind is the creation of value." That is, a product or service that offers clients greater value than what it costs to manufacture and deliver. Consequently, the pricing process begins with acquiring a good understanding of customers' wants, needs, and desires. The internet offers a means to encourage feedback from customers. An efficient pricing policy begins before R&D. According to Simon and Schumann, the internet significantly increases the possibility of creating value for the client, in three primary ways: it increases the exchange of information between buyers and sellers; it increases the company's capacity to segment the market and to offer mass customization; it significantly reduces the translation costs; and it facilitates doing business.
2. The internet adds value by giving the customer the possibility to make an order and check its progress in a simple way. Creating value for the client can translate into higher prices.
3. As well as increasing the capacity to segment the market, the internet also increases the capacity to differentiate prices (Baye et al. 2007). Given that segments have different degrees of elasticity with respect to price, pricing formulas can vary from segment to segment. As customers provide more and more information, best price estimates for the segment can be updated continuously. Moreover, with its interactive communication structure, the internet is a platform that can sustain a mass customization programme.
4. Unless the manufacturer is capable of always offering the lowest cost, a price war needs to be avoided. To manufacturers that drop prices, competitors can respond by lowering theirs in turn. The answer cannot be a new price cut. Customer loyalty would plummet. The best response is to offer more reasons to select the product they are offering, at a fair price. "More value for money, not the lowest price". Flexibility and simplicity in the pricing policy, for example, can be aspects that create value for the client, just like the possibility of using the internet to make the pricing structure more transparent and to constantly update the options on offer.
5. The marketing strategy cannot be based on the assumption that pricing structures and competition structures are stable. The development of digital technologies in

processing information flows is exponential. Both buyers and sellers benefit from it, but to different extents, hence equilibriums change quickly. We cannot expect stability in the environmental aspects that affect prices (i.e., rate of exchange, interest rate). The successful firms are those that are able to adapt constantly to an ever-changing environment.

Digital technologies have increased the possibility of better understanding how clients respond to policies geared at getting certain prices accepted by offering other elements in exchange, such as simplicity and time-saving in the configuration of the desired car and flexibility in the modes of use. This is especially true for vehicles offered to high-end segments. It is the case with the "all-inclusive" lease package introduced by Lexus for the new UX product.[4]

Place. At the beginning of the 2000s, Tapscott observed that "Every b-web competes in two worlds": that of the physical product (marketplace) and that of the digital world of information (marketspace). This author had anticipated that, "Within a decade, the majority of products and services in many developed countries will be sold in the marketspace." He also wrote that, "a new frontier of commerce is marketface," the interface between the marketplace and the marketspace. He referenced Gap as an example of the "aggregation of online presence and physical stores." Gap sold online (marketspace). If the client then found that the product was not the right size, they could return it to a store on the chain (marketplace).

While in some sectors the internet has rendered physical sales structures unnecessary, in the automotive industry dealerships are irreplaceable for most activities: showroom management, routine maintenance and controls, and after-sales services. In almost every country, franchisees have obtained state legislation to protect their position. Given that many car buyers prefer searching for vehicles online and making their decisions before they even consider visiting a dealership, manufacturers make increasing use of augmented reality (AR) and virtual reality (VR) to present the features of their products to potential customers, and wish to maintain direct contact with prospective customers.

Car manufacturers do not give up on innovating when it comes to distribution. Ford, for example, has used the world's largest vehicle market, Cina, as a test for alternatives to traditional dealer showrooms. It reached an agreement to manage the new initiative with Alibaba's online retail platform, Tmall. According to the analysts, the attempt at alternative sales models makes sense in China, because the market is highly competitive, growing and, much like the U.S., consumers are increasingly interested in online sales.

---

[4]"Lexus wouldn't want to make the short-term lease program too appealing in price". "We want to know what's the threshold for this" and "It has to be priced right." Lexus is pitching the UX at young, urban customers, or people with city-centered lifestyles, touting its tight turning radius; standard Apple CarPlay; safety features that can detect and brake for pedestrians or cyclists; and a mobile app that brings Amazon's Alexa voice-controlled capabilities into the vehicle. Source: "Lexus' UX will test market for flexible lease options," *Automotive News*, 14 October 2018.

New relationships between manufacturers and dealers. The prospect of an automotive market with growing EV sales made it necessary to rethink the relationships between manufacturers and dealers. For example, an agreement between VW and its European dealer network rewrote the way in which VW brand retailers make profits, and the relationships they have with customers and factories. The agreement allows the automaker to develop direct links with their customers, "But it includes revenue-sharing arrangements and eases infrastructure demands." The aim is to guarantee dealers a given level of profitability, given that they will sell EVs, which will generate less demand for repair work and other services. The European agreement allows for direct sales of vehicles to consumers over the internet, though VW expects them to make up 5% at most.[5]

Not everyone believes in a dealership crisis. Despite the arrival of electrified, autonomous cars and the fears of growing online sales, they are convinced that dealerships still have a future. For example, the CEO of Audi America, Scott Keogh, defines the change underway as "a massive opportunity for dealerships." Four components of dealership revenue—new-vehicle sales, second hand-vehicle sales, services and parts, and new digital services—will provide benefits, Keogh affirmed, arguing that those that anticipate disasters are wrong.[6]

Progress in complementary structures too. Structures that are complementary to distribution channels, such as petrol stations, also show developments, with a succession of identifiable stages, in response to customers' expectations, which in the car industry change at almost regular intervals, as demonstrated by research conducted by D'Aveni (2000). "The gasoline stations have redefined the primary benefit they offer at regular ten-year intervals to serve different needs," driven by variations in consumer needs originating from changing life patterns and technological advances. D'Aveni predicts that they will become "greener centres" for alternative fuels, such as "electricity for hybrid cars and water for hydrogen cars" (see Table 1).

Towards internet-based multichannel distribution. In the current structure, manufacturers do not usually have direct contact with clients. Only a few own a flagship

---

[5]The president of Volkswagen's European Dealer Council, who played a key role in the European renegotiations, helping to define new contractual obligations for dealerships and the factory, said that with the coming EVs and digitalization of retail in Europe, VW's European network had reached an inflection point. Either its longstanding factory-dealer business relationship had to change, with dealerships learning to live on lower revenue, or the stores would face a financial crisis. The agreement, which came into force in April 2020, put an end to the manufacturer's request for costly "glass palaces," replaced with simpler, more cost-efficient dealerships. Source: VW rewrites the rules with new retail agreement, *Automotive news*, 29 October 2018.

[6]It's not the end of the world, Keogh, 48, said at the Automotive News World Congress. "It's the beginning of a powerful, all-new world." Likewise, Porsche Cars North America CEO Klaus Zellmer has his eye on how dealerships can elevate the brand's customer experience to a level to rival luxury brands both in and out of the auto industry. Though the purchase process will transform—Porsche expects 30% of all its cars will be sold online by 2025—he insists the dealership will remain a key element. Source: "Dealers are vital in electrified, self-driving future," *Automotive News*, October 2018.

Table 1 As technologies changed, the offering of gas stations changed too (in the U.S.)

In the 1960s	They were *full-service stations*, offering mechanics, window washing, and high-performance gas to serve the needs of muscle cars, and the desire for personal service from a trusted source
In the 1970s	They became *self-service filling stations*, offering do-it-yourself pumps with choices of unleaded and diesel fuel in many new locations to serve the needs created by the energy crisis, a more on-the-go lifestyle, as well as offering the assurance of not being stranded without gas
In the 1980s	They become *convenience stores* using gas and automated cars washes as lures to bring people into a retail store. The goals were to serve the needs of people whose longer work days and commutes created the need for fast one-stop shopping
In the 1990s	They became *safe and secure havens*, serving the needs of lone women for improved safety and security during extended hours by offering faster ways to pay (e.g., speed passes and credit card pumps) coupled with better lighting and security cameras for night-time purchases
In the 2000s	They converted their retailing from small convenience store selections to a much broader array of merchandise and services, including an alliance with Dunkin' Donuts and high-end coffee retailers
In the future	Many believe gas stations will morph into green centres for alternative fuels, such as electricity for hybrid cars and water for hydrogen cars

Source Analysis and synthesis of information taken from D'Aveni (2000), pp. 96–99

store in the major cities. Manufacturers provide dealers support with marketing, sales, after-sales, and services, but it is the dealer that has direct contact with the client. In future, the structure will change drastically given that digital transformation is opening up new segments, such as mobility services, connected services, and intermodal transport (see Chap. 18). The sales structure therefore needs to change so that the client can be at once a car owner and service client. The current structures are very varied, which complicates the shift towards digitalisation. Another difficulty is that customers expect the same quality of service, whatever the channel.

According to Winkelhake (2017), "In order to be successful, the established manufacturers must transform the sales structure into a 'multichannel' structure with many customer accesses." He also indicates a series of trends and changes in the sales department that need to be taken into account when rethinking the sales structure due to digital transformation:

- Internet sales of spare parts increase. 3D printing of spare parts increases.
- The number of dealers decreases significantly. Of those that remain, the successful ones will be those that are part of organisations or commercial chains, and which extend their activities to include new services (including complementary ones such as financing, insurance, and others).
- Virtual Reality and Augmented Reality become increasingly important in the sales process, the configuration of the desired vehicle in built-to-order models, and in virtual test drives.

- In mobility services, the manufacturing brand loses importance (see next chapter). In these services, the customer's choice criterion is based on price and the degree of extension of networks in the territory. Loyalty to mobility platforms is achieved with commercial models and customer programmes.
- It is anticipated that of the average dealer's earnings, those deriving from car sales will drop, being partly replaced with those deriving from mobility service sales, continuous vehicle diagnoses, maintenance work based on cognitive technologies, and software updates "over the air".
- Organisations independent from manufacturers will enter the market, who manage internet-based sales on multibrand platforms.

Promotion. The digital age has changed the way in which people and businesses communicate. It has opened up new communication channels, particularly those for communicating the benefits of a product to clients and assisting in buying decisions. Advertising, promotion, publicity, public relations, and most other aspects of corporate communications are partly outdated concepts. Some even speak of "the end of an era." Until recently, manufacturers communicated in a one-way, one-to-many manner, sending messages to a "powerless consumer." In the digital age, the relationship was reversed. Now, customers have access to a great deal of information about products, and "power shifts toward them." They do not obtain information only from the manufacturer or from intermediaries as in the past, but also from other consumers "one to another" via the web (Adjei et al. 2010). What the customer perceives and considers depends ever less on what the company communicates about itself.

In *The Art of Digital Marketing* (2016), Dodson observes that, unlike traditional communication channels, the internet allows companies to understand a lot about their clients. By looking at what people do online, marketing management can understand various aspects of their buying behaviour: from preferences regarding the technology they use to internet usage habits and what they want to know before formally initiating a buying process. Capacities to better target products are thus increased, as are the possibilities of building successful marketing strategies.

Table 2 sums up the main elements of the promotion mix, and indicates some means for their online implementation.

What are the distinctive features of digital interactive media? In "*Digital Marketing Communication*" (2001), Deighton and Barwise responded to this question by identifying the following features:

- *Any-to-any, not one-to-many communication.* The internet allows participants to connect with one another. Consumers can form communities that exchange information. This gives them more power and a greater capacity to evaluate communications made by companies, a power that traditional media forms did not provide.
- *Content can be perpetually fresh.* Digital interactive media can be updated continuously, which is an advantage that brochures, catalogues, and other traditional forms did not have. Prices can be revised in line with supply and demand.
- *Consumers can select information.* Search and indexing engines can select the enormous mass of data contained on web pages.

Table 2 The main elements of the promotion mix

Communication tools	Online implementation
Advertising	Interactive display ads, pay-per click search advertising, social networks, viral advertising campaign
Selling	Virtual sales staff, site merchandising, chat and affiliate marketing, augmented reality, avatar[a]
Sales promotion	Incentives, rewards, online loyalty scheme, special series, geo-localizer
Public relations	Online editorial, blogs, feeds, e-newsletter, newsletter, social networks, links and viral campaigns, influencers, bloggers
Sponsorship	Sponsoring an online event, site or service, influencers
Direct mail	Opt-in email using e-newsletters and e-blasts (solus e-mail), GPS tracking (smartphone)
Exhibitions	Virtual exhibitions and white-paper distribution
Merchandising	Promotional ad-serving on retail sites, personalised recommendations and e-alert
Packaging	Virtual tours, real packaging in displayed online
Word of mouth	Social, viral, affiliate marketing, e-mail a friend, links, online forum, influencers, bloggers

Source Analysis of documentation taken from Chaffey and Chadwick (2016) and fitting to the industry
[a] An avatar is a graphic representation that can be animated by means of computer technology. Using an avatar sales agent leads to more satisfaction with the retailer, a more positive attitude toward the product, and a greater purchase intention" (Holzwarth et al. 2006, pp. 19).

- *Communities can form, unbounded by space or time.* Consumers have always had the possibility of forming communities and exchanging information, but digital interactive media make it possible to set them up more quickly and to extend them to broader geographical areas.

According to Chaffey and Chadwick (2016), there are three ways of applying digital technologies to the various elements of the promotion mix. The first is "reviewing new ways of applying each element of the promotion mix" (indicated in the previous table). The second is identifying how the internet can be used in the various phases of the buying process. Lastly, the third involves using communication tools (also indicated in Table 2) "to assist in different stages of consumer relationship management," from customer acquisition to retention. In a web context, this means attracting initial visitors to the site and encouraging repeat visits through the various communication tools.

People. The fifth element of the marketing mix consists of the interactions that develop in the pre- and post-sales stages between members of the organisation and clients. It concerns how the organisation provides services to clients. This primarily occurs in the dealership and stores owned by the manufacturer. In the automotive industry, staff have a high level of contact with customers. For this reason, it is important for the service organisation (dealerships and others) to clearly specify what is expected from personnel in their interactions with customers. "To achieve the specified standard, methods of recruiting, training, motivating and rewarding staff cannot be regarded as pure personnel decisions, they are important marketing mix decisions" (Palmer 2014). "People" also underscores that people are key players in the search for information, not merely passive listeners of companies' messages.

Process. In the marketing mix, process concerns the methods and procedures that the company employs to carry out all the marketing functions, in particular those that directly involve the client. It primarily concerns promotion and customer services. When the organisation produces a service (i.e., information that precedes the car purchase), the customer becomes part of the process itself. They "co-produce" the service. Is it the first time they are buying a car or the third? Are they passionate about cars, or do they simply see them as a means of transport? Are they sensitive to environmental protection issues?

Physical evidence. In an online context, physical evidence is what the client perceives in terms of navigation, availability, and performance. It is the customer's experience of the company through the website. When, as is increasingly common, the offer of a mechanical product such as the car is combined with new services, principles need to be applied to be successful. The intangible nature of the service means that potential consumers cannot evaluate it before buying, which increases the risk for them inherent in any buying decision. Marketing must therefore intervene to reduce the level of risk, offering tangible evidence of the characteristics of the service. In its simplest form, it could be a video with photographs to illustrate the place and the equipment with which routine checks and maintenance will be carried out.

From AIDA to five 'A' stages

In *Marketing 4.0*, Kotler et al. (2017) made an important contribution by identifying the critical points in the passage from traditional marketing to digital marketing. In particular, dealing with the effects of digital technology on the consumer buying process, the authors propose a step forward compared to the traditional path of AIDA: *attention, interest, desire, and action*. Taking into account the changes framed by connectivity, the new customer path is: *aware, appeal, ask, act, and advocate*.

(1) "Aware." Clients are passively exposed to company communications, receive communications by word-of-mouth from other clients, and draw upon their own past experiences. (2) "Appeal." Clients develop the messages to which they have been exposed. They focus their attention on a limited number of brands. (3) "Ask." They seek information and advice within the family, from friends, the media, or directly from manufacturers' websites or sales agents. (4) "Act." Convinced by the information at their disposal, clients purchase a given brand. They experience purchasing and post-sales services. (5) "Advocate." If their experience in using the product confirms the validity of their choice, this creates trust, which may lead to repurchasing and advocacy (word-of-mouth), by means of which clients communicate their positive experience to others.

At the end of the process, Kotler et al. conclude, *Marketing 4.0* aims to lead customers from awareness to advocacy. To influence this transition, marketers act in three ways, combining "their *own* influence, *others'* influence and *outer* influence". Their "own" influence comes from within the organisation, "others'" influence comes from a community to which the customer belongs, and "outer" influence comes from external sources such as advertising and other marketing stimuli.

References

Adjei MT, Noble SM, Noble CH (2010) The influence of C2C communication in online brand communities on consumer purchase behaviuor. J Acad Mark Sci 38(5):634–654

Baye MR, Gatti JRJ, Kattuman P, Morgan J (2007) A dashboard for online pricing. Calif Manag Rev 50(1):202–216

Burke R, Rangaswamy A, Gupta S (2001) Rethinking market research in the digital world. In: Mahajan V, Wind J (eds) Digital marketing: global strategies from the worlds' leading experts. Wiley, London, pp 226–255

Canzler W, Knie A (2016) Mobility in the age of digital modernity: why the private car is losing its significance, intermodal transport is winning and why digitalisation is the key. Appl Mobilities 1(1):56–67

Chaffey D, Chadwick E (2016) Digital marketing. Pearson

Currid-Halkett E (2017) The sum of small things: a theory of the aspirational class. Princeton University Press

D'Aveni R (2000) Beating the commodity. Harvard Business Press

Deighton J, Barwise P (2001) Digital marketing communication. In: Mahajan V, Wind J (eds) Digital marketing: global strategies from the worlds' leading experts. Wiley, London, pp 239–261

Dodson I (2016) The art of digital marketing. Wiley, London

Holzwarth M, Janiszewski C, Neumann MM (2006) The influence of avatars on online consumer shopping behavior. J Mark 70(4):19–36

JD Power and Associated (2000) More than half of all new vehicle buyers use the internet during the shopping process (press release, September 14)

Kotler P, Kartajaya H, Setiawan I (2017) Marketing 4.0. Moving from traditional marketing to digital. Wiley, London

Oge M (2013) Driving the future: combating climate change with cleaner, smarter cars. Arcade Publishing

Palmer A (2014) Principles of service marketing. McGraw-Hill, NY

Ryan D (2015) Understanding digital marketing: marketing strategies for engaging the digital generation. Kogan Page

Simon H, Schumann H (2001) Pricing opportunities in the digital age. In: Mahajan V, Wind J (eds) Digital marketing: global strategies from the worlds' leading experts. Wiley, London, pp 362–390

Sumantran W, Fine C, Gonsalvez D (2017) Faster, smarter, greener: the future of the car and urban mobility. The MIT Press

Tapscott D (2000) Digital capital: harnessing the power of business webs. Harvard Business School Press

Veblen T (1899, new ed 1924) The theory of the leisure class. Allen & Uniwin, London

Winkelhake U (2017) The digital transformation of the automotive industry. Springer, Berlin

Chapter 18
Innovation and Digital Transformation in the Automotive Industry

Abstract On the eve of the 2020s, the concept of "digital transformation" is discussed increasingly frequently in the car industry. It has brought about a profound change in the competition in all sectors of the economy, so this chapter examine how significant its effects are in various sectors. Analyzing the current situation regarding specific trends in the automotive industry, the aim is to offer a prediction as to the evolution of marketing strategies as we move towards the 2030s. Three forces, acting together, created the new digital world. The first, in order of time, was the exponential growth and ever lower costs of computer power. The second force is the value of networks, which has grown as they have grown in size, and the third is that more data have been transmitted at an ever lower cost using cloud computing technology. In car industry political, environmental, social, and economic trends are changing the competition. Faced with technological change occurring at an exponential rate, companies are much slower. Filling or reducing the gap is the main challenge faced by management. Furthermore, in the automotive industry four innovative trends merit particular attention thanks to their likely rapid evolution over coming years and their impact on the entire industry: mobility services instead of vehicle ownership; increasing demand for connected services; autonomous driving; and electromobility (EV). The chapter tries to answer to the following question: Why, even though it is difficult to predict the consistency of EV and driverless cars demand, are almost all manufacturers investing significant amounts?

1 The Digital Transformation is Accelerating and Will Leave Many Behind

"Digital transformation is the adaptation a company makes to be competitive in a digitized world". That is the definition given by Jansson and Andervin in their book *Leading Digital Transformation* (2018). On the eve of the 2020s, this concept is discussed increasingly frequently in the car industry. Digital transformation has brought about a profound change in the competition in all sectors of the economy. One merely needs to examine certain trends to see how significant its effects are.

© Springer Nature Switzerland AG 2019

E. Candelo, *Marketing Innovations in the Automotive Industry*, International Series in Advanced Management Studies, https://doi.org/10.1007/978-3-030-15999-3_18

- In 2006, the first six companies among the Fortune 500 in terms of capitalisation were Exxon Mobile (oil), General Electric (conglomerate), Microsoft (tech), Citigroup (financial services), British Petroleum (oil), and Royal Dutch Shell. Twelve years later (2018), the list was dominated by technology businesses. In order, in the top five positions were Apple, Amazon, Alphabet (holding company of Google), Microsoft, and Facebook. Exxon Mobile was demoted to eighth position, preceded by Berkshire Hathaway (financial) in sixth place and JPMorgan & Chase (financial) in seventh.
- Of the 61 companies included in Fortune's original list, only 12% were still on the list in 2017.
- Foster and Kaplan (2011) calculated that the average life of companies in the S&P Index 500 went down from 61 years in 1958 to around fifteen years in 2017, and they estimated that three quarters of the companies that would be listed on the S&P Index 500 had not yet been established in 2017.
- In twenty years, Amazon went from being an online bookstore to a global e-commerce retailer, surpassing, in 2018, together with Apple, a trillion dollars of market capitalization, a level never reached by others. In 2017, Amazon became a potential competitor to third-party auto shopping guides when it unveiled its new Amazon Vehicles research hub.
- Tesla overtook General Motors in terms of market capitalisation despite the ratio of production volumes between the two companies being 1 to 100. At the end of 2017 Tesla had €235,000 of market capitalisation for every car sold, while the figure for VW was €6300.

What does all this mean? It means that no company, however successful it has been, can rely on existing strategies. It also means that digital technology is critically important in every sector, and the car industry is no exception to this rule.

At the Davos 2016 Congress, Mary Barra, CEO of GM, said "We are moving from an industry that, for 100 years, has relied on vehicles that are stand-alone mechanically controlled and petroleum fuelled, to ones that will soon be interconnected, electronically controlled and fuelled by a range of energy sources." "I believe that the auto industry will change more in the next 10 years than it has in the last 50."

Venkatraman (2017) summed up the opinion of many scholars when he wrote, in *The Digital Matrix*, that we are only in the first stage of the digital transformation, and we do not know what the future holds. He added that: "We see companies born digital in the post-industrial age emerge with principles and practice of management that are very different from the companies born before them".

2   The Key Challenge is the Rate of Change

In the following pages, we will assemble the opinions of experts, car manufacturer managers, consultancy firms specialising in the automotive industry, and authors of articles and books on the subject. The aim is to analyse the current situation and to offer a prediction as to the evolution of marketing strategies as we move towards the 2030s.

In the last two changes, we described the medium-term trends underway. Given that the forecasts, despite being a long way off, are still affected by the situation in which they were formulated, it is useful to recall the context leading up to the 2030s.

"iPhone moment". These two words have become emblematic of the revolution that digital technology has brought about in various sectors, such as that of telephone services, where it has led to the decline of icons like Nokia and Blackberry. According to many experts, for the car industry this moment is still far off and could be warded off with appropriate strategies. The management of the major car companies sees electric and self-driving vehicles on the horizon, connected to the web and shared among users rather than bought separately, and is preparing itself accordingly.

The worst forecasts for the industry are based on a possible future combination of electric cars, driverless cars, and ride hailing. "Combine all three, for example through an Alphabet investment in Lyft, and you have a new model of transport as a service that would be a cheap compelling alternative to traditional car ownership", according to RethinkX, a think tank that analyses technology-driven disruption. "The smartphone and its apps made new business models possible," said Tony Seba, a Stanford University economist "The mix of sharing, electric and driverless cars could disrupt everything from parking to insurance, oil demand and retail." Ghosn, CEO of Nissan-Renault at the time, holds the same opinion, having urged the industry to embrace innovative newcomers. "Three forces - electrification, autonomous drive and connectivity - are about to change our industry in ways we are only beginning to imagine", he said.

The main elements of the car industry on the eve of the 2020s are as follow:

- Market valuations of car manufacturers are at their lowest level since the financial crisis (2008), due to scepticism that the current level of profitability can be maintained. Furthermore, the slowdown in demand, for the first time since World War II, is considered to be structural.
- In the EU, new vehicle registration plummeted heavily when new emission standards took effect. From London to Prague, diesel bans and driving restrictions in most global cities have been introduced or threatened.
- Manufacturers invest considerable sums in battery and autonomous driving while the appeal of brands declines as clients, especially the new generations, pay more attention to dashboards than horsepowers. It is difficult to maintain high brand loyalty.
- The car industry proves vulnerable to political events. Protectionism is bringing new uncertainties in an industry heavily reliant on global supply chains. The trade war between the U.S. and China has unsettled the expectations of many choices regarding the geographic distribution of supply chains and overseas investments. An example can be found in the decision by German manufacturers to invest in U.S. plants to supply both the internal market and the Asian market.

- The shrinking car industry in the U.S. (Detroit) and in Italy is a warning sign. Nations that still have high production levels, such as Germany (over 5 million vehicles per year), know that their fate is tied to that of the internal combustion engine (ICE).
- The frame of electric vehicles is profoundly different from that powered by a combustion engine. Were the demand for EVs to accelerate quickly, the supply chain for the car industry would be unsettled primarily in Europe. Producers of mechanical parts cannot easily transition to producing components for an EV.
- China is already the leading market for EVs by a long way. The history of the car industry teaches us that pioneers can quickly consolidate their position and dominate the market, especially if they operate with large volumes and expand continuously. Being fast second in a radical shift in technology (from the ICE to EVs) is very difficult.
- Bloomberg estimates that Europe controls a mere 4% of the production of batteries, while the EV supply chain is in Asia, and China controls it.
- If the transition to the EV occurs within decades, car manufacturers will have time to adjust. However, in the short-medium term (3–5 years), the response will be tough. Production volumes are low and only premium brands earn enough to finance the change. The major carmakers are investing in EVs (the Germans have more than 30 models for sale), but they are far from breaking even. The most common response is alliances. VM is partnering up with Microsoft and Gett, BMW with Intel and Mobileye, Daimler with Bosch and Uber, Renault, Nissan, and Mitsubishi with Google.
- The history of the car industry in Europe teaches us that given the importance of this industry (Jullien and Pardi 2013) for employment, in the face of a prolonged fall in demand governments intervene with various forms of bailouts. The result is often an excessive production capacity and consequent downward pressure of prices and investments.

Avoid a Nokia-style downfall. Three forces, acting together, created the new digital world. The first, in order of time, was the exponential growth and ever lower costs of computer power ("Moore Law"). In an article released in 1965, Moore (1965) observed and described how at a fixed time (at intervals of two years), the performance of computer chips continued to grow at an exponential rate, while the price and size of chips dropped at a corresponding asymptotic rate. Given the continuous changes in technology, the interval was repeatedly adjusted. Currently, it is between 18 and 24 months. The "Moore Law" is an observation; it is not scientifically proven but it is taken as a standard of the digital revolution in industry. The second force is the value of networks, which has grown as they have grown in size ("Metcalfe's Law"), and the third is that more data have been transmitted at an ever lower cost using cloud computing technology ("Gilder's Law"). As Wadhwa and Salkever point out in their book *The Driver in the Driverless Car*, (2017) "Moore's Law" explains why the iPhone and Adroid phone are much faster than the supercomputers of decades ago, and even faster than the computers that NASA used to send a man to the moon during the Apollo mission.

Faced with technological change occurring at an exponential rate, companies are much slower. Filling or reducing the gap is the main challenge faced by management. For marketing, the task is even more difficult. It must tackle the problem posed by the difference in speed in the rate of change of the environment (not only technological, but also social), and the rate of change that, in order to adapt, the organisation can undertake and support internally. External change is stronger, while that of organisations is considerably slower, held back by the inertia created by the organisational structures, attitudes, behaviours and cultures that slow down decisions, process adaptation, the identification of imminent threats, and understanding how value (profitability) is migrating along the value chain. Even in the social context change is quicker; one need merely think of the new attitude of young people in relation to owning goods ("Use, don't own").

Digital transformation heightens the difficulties of marketing in avoiding the investment of resources leading to solutions destined to become obsolete in the short term. The experience of car phones and consumer GPS units is informative. Car phones became very popular in the 1980s and 1990s, but quickly disappeared as a result of the diffusion of mobile phones. Consumer GPS units, like TomTom or Garmin, which became part of the equipment of many cars, had replaced road maps. But because nearly everyone now has a smartphone with several mapping apps, sales of consumer mapping devices have plummeted.

Marketing and technology are closely tied. The major advances in marketing were the result of technological innovations. Printing, the telegraph, railways, radio, television and, recently, the internet, have radically changed the way of communicating with consumers and how consumers can communicate among themselves.

In *Understanding Digital Marketing* (2015), Ryan, after writing that the internet "heralds the single most disruptive development in the history of marketing", points out that, driven by a new technology, the process of marketing development follows four phases: (1) A new technology emerges. It is initially used only by experts or technology enthusiasts, or those attracted by innovations; (2) The new technology begins to spread and to occupy a place on the market. It is adopted in the marketing of many firms; (3) The companies most open to innovation in marketing explore the new technology to reach their target of prospective consumers; (4) The technology begins to be commonly used. This surprises those that do not adopt it in marketing. When the potential of the new technology is fully exploited, it has the power to open up new markets and to shake the foundations of existing ones.

3 Unexpected Trends That Are Shaping the Car Industry

Four trends merit particular attention thanks to their likely rapid evolution over coming years and their impact on consumer behaviour and on the entire industry: mobility services instead of vehicle ownership; increasing demand for connected services;

autonomous driving; and electromobility (Firnkorn and Müller 2012; Winkelhake 2017; Jansson and Andervin 2018; Sumantran et al. 2017). Given their significant impact on industry innovation, autonomous driving and electromobility will be analysed in specific sections.

Mobility services instead of vehicle ownership

In emerging countries and outside of large cities in the more advanced economically countries, car ownership still holds a leading position in the scale of aspirations (it is often in second place, after home ownership). This is not the case in major European cities, cities in the United States, and the megacities in other nations. In these areas, the impact of car traffic on the environment and time wasted as a result of traffic congestion reduce the desire for car ownership, especially among young people, and the demand for mobility services increases, in turn, facilitated by technological advances (such as the use of apps). This growth will continue in coming years, driven by the dissemination of electrically driven autonomous vehicles. In the future, a "robotaxi" could be hailed through a smartphone app, be ready at the place indicated to take the client to their destination, collect payment online, and be guided through traffic using a remote control system.

> Toyota was one of the first automotive companies to confront the change under-way. It sought to reposition itself in the automotive industry and acquire advantages over its competitors in a time of uncertainty in which automated driving and the sharing economy threaten to displace the traditional model of vehicle ownership. It has invested both in Grab (ride-hailing) and in Uber Technologies (robotaxi).[1]

The mobility services offer is dominated by a few: Uber, Lyft, Zipcar, BlaBlaCar, and Didi. They offer very similar services distinguished by just a few details. The success of these companies can be attributed to their being easy to use for clients and low-priced. Established companies cannot compete with these companies, which are streamlined, do not have assets (or have few), and above all are "born in the web".

Some traditional manufacturers offer mobility services, often in cooperation with partners, e.g., Daimler and BMW. Some offer intermodal services to clients that cover long distances using different modes of transport—car, bus, bike, ferry, i.e., Daimler with Moovel (see below). Ford bought out Spin, an electric scooter company, in a bid to expand its mobility offerings and reach consumers needing short-distance transport solutions. In the U.S., VW introduced its division Moia, with services including ride-hailing and car-sharing. Moia differs slightly from the likes of General Motors, BMW,

---

[1]Toyota is preparing for a potential future where people don't buy cars. That's behind the hefty investments that the company has made in ride-hailing providers, most prominently the $1 billion that it poured into Southeast Asian leader Grab. With Uber Technologies Inc.—into which Toyota poured half a billion dollars—the automaker is designing a specialized minivan for their robotaxi project. "Toyota, plowing millions into Uber, eyes the future of cars", *Automotive News*, 28 September 2018.

and Daimler in that it aims to compete with established public transportation. For that reason, the service uses custom-made Moia-branded electric vehicles with six to eight seats offered to those who want to share their ride.

To avoid or reduce conflicts between those selling vehicles and those selling mobility, some manufacturers have created a distinct organizational division. Daimler, for example, has set up the Moovel division, which offers a range of services and has multiple partners: car-sharing under the Car2go brand (also of Daimler), public transport, taxis, and bicycle services. To better compete with the U.S.-based ride-hailing service Uber, Daimler and BMW have combined their car-sharing services Car2Go and Drive Now. "As pioneers in automotive engineering, we will not leave the task of shaping future urban mobility to others", Daimler CEO Dieter Zetsche said when the partnership was announced.

What is the forecast for carmakers? It is plausible to predict that the trend towards mobility services will have a negative impact on the sales volumes of car manufacturers in that it will reduce brand loyalty. Clients of mobility services are interested in the simplest, most flexible mode of transport. The brand of car manufacturers therefore becomes less important. Moreover, car-sharing becomes more efficient thanks to its greater use compared to that of a private vehicle (which remains stationary for longer).

In the mobility services offer, brand-independent platform operators could establish themselves. Their offer could in fact prove considerably broader than that of car manufacturers, given that it could contain more brands and more types of vehicles, and therefore be more attractive to customers. For these operators, volume brands are practically equivalent. The choice of supplier is based on price, especially if used for robotaxis.

Following the increase in vehicles used for mobility services, the new car market will contract in megacities, combined with the lower interest in private car ownership and the spread of autonomous driving in megacities themselves. The consequence that many predict is a fall or consistent slow-down in the demand for vehicles in more economically advanced countries, which should nonetheless be compensated by a greater demand in emerging markets.

Increasing demand for connected services.
The number of new cars equipped with connectivity features is destined to rise in the years to come, again because of technological advances. There will be two main consequences. On the one hand, connectivity facilitates the use of mobility services and therefore expands the demand for them, and on the other hand it gives manufacturers greater possibilities for new business models (i.e., remote diagnostics). Given that the number of sensors and cameras required to provide driving assistance leading to autonomous driving is increasing, data are also becoming increasingly available. These data can be connected and evaluated with those deriving from knowledge of the environment in which the vehicle finds itself, and with connected services. As such, it is possible to obtain useful information about customer behaviour.

According to some experts in the industry, the trend towards greater connectivity necessarily leads to collaboration between multiple operators, because the consumer wants continuity between one service and another, and above all fast response times.[2]

4   The Road Towards Autonomous Driving

When, in 2014, Google presented a fully autonomously driven car without pedals or a steering wheel, a new phase in the evolution of the automotive industry began. From then on, among manufacturers no one doubted that the driverless car was the future. As always, as the breaking point draws near, the problem for manufacturers or suppliers is predicting *when and how* it will occur.

One of the few certainties is that the road towards autonomous driving will take place in evolutionary steps. To indicate the stages from total human control to total car control, the standard classification of the Society of Automotive Engineers (SAE) is generally used. In the first three stages, control of the vehicle is in the hands of the driver, assisted by autonomous systems such as in level 2, when the keeping land and distance systems intervene (see Table 1). It should be pointed out that assistance systems (levels 1 and 2) are a powerful marketing tool because drivers see them as elements that offer greater driving safety and are prepared to pay for them. From levels 3 to 5, the evolution ranges from highly automated to complete automation. The expressions "self-driving", "autonomous" and "driverless" are often used interchangeably, creating confusion. Simoudis (2017) clarified the meaning of these expressions, as summarised below.

All the leading manufacturers are experimenting with autonomous driving. Approaches differ, as do the horizons placed on the achievement of fully autonomous driving. For example, in 2016 Ford announced the plan to produce a fully autonomous vehicle within five years, while GM opted for a gradual approach to self-driving cars. This began with a semi-autonomous system called Super Cruise, which takes only the most routine driving away from drivers. GM technicians predicted that it could take around ten years to perfect autonomous driving. Table 2 summarizes some significant examples.

In the medium–long term, technical difficulties, legislative interventions, and legal problems that will slow down the evolution towards level 5 could be anticipated.

---

[2]Who is driving the connectivity agenda? To Agustin Martin, Toyota VP New Mobility and Connected Car; two questions were asked: (1) Who is driving the connectivity agenda? Is it mostly coming from consumers or manufacturers? The response was: Consumers. There is a clear desire on their part to have a seamless life. We must respond to this and technology is the enabler. (2) What is the biggest challenge for connectivity and how will you address it? Response: Adjusting a 100-year-old planning process to the "I want it now" mindset of today's consumers. We will do this through wider collaborations than ever before, both inter and intra industry as we are both trying to satisfy the same consumer. This seamless life that consumers want requires many operators and industries to talk to each other and this makes it one of the most exciting moments in our history. From *Global Monthly*, September 2018.

Table 1 Levels of driving automation

Level 0	Refers to *no automation*. The human driver has control of every aspect of the vehicle at all times
Level 1	Vehicles include driver-assistance functionalities that are used for steering, braking, and acceleration using data about the driving environment. The driver is still responsible for monitoring the driving environment and controlling most vehicle features
Level 2	In addition to braking and acceleration, the driving system becomes responsible for lane centring, implying that the driver can take their hands off the steering wheel and their foot off the pedal
Level 3	At this level, all of the vehicle's safety functions are taken over by the driving system. The vehicle can be characterised as *self-driving*. However, it is expected that the human driver will always be able to intervene and take over when necessary
Level 4	Vehicles at this level are *autonomous*. They can take over vehicle functions and monitor road conditions and overall environment conditions for an entire trip, even if the human driver does not respond to a request to intervene and take over when asked
Level 5	Level 5 refers to *complete automation*, where every aspect of the vehicle is controlled by the driving system (software, hardware, data), the so-called *driverless vehicle*, such as the experimental "Google Cars"

Source Synthesis and analysis of information adapted from Simoudis (2017, pp. 30, 31)

Table 2 The road to the self-driving car (some examples)

Level 0	Ford Model T (1908). No automation
Level 1	Jaguar XK with Adaptive Cruise Control system (1996). Adaptive Cruise Control speeds the car up and slows it down to maintain a preset distance from the vehicle in front
Level 2	Mercedes-Benz S-Class with Active Lane Keeping Assist system (2013). If the car wrongly crosses road lane margins, Active Lane Keeping Assist takes control of the brakes to rectify the error
Level 3	GN Cadillac CT6 with Super Cruise system (2017). Super Cruise should take control of the car on motorways, but can hand control back to the driver
Level 4	Ford Fusion with autonomous system (due on the market in the 2020s). The autonomous system should be able to handle entire journeys without any human intervention
Level 5	Google self-driving car (possibly on the market by the end of the 2010s). Designed to be fully autonomous in all situations with no steering wheel or pedals

- To bring vehicles with an increasing level of driving automation to the market, innovation is required in four areas: hardware technology, software technology, broadband connectivity, and mapping. Driverless vehicles need very detailed 3D maps (with precision to the centimetre) to "understand" the environment. Such maps generate large quantities of data, up to a tetrabyte per day. To date, there has been no agreement between map companies on standards or data sharing. To reach level 5 driving automation, several new technology innovations and extensive additional testing in various environmental conditions will be necessary to convince the authorities and consumers that robocars can be trusted.

- Legislative interventions. On the eve of 2020, only a few companies have authorised the testing of driverless cars on their territories.
- Legal problems. In the event of an accident, who is responsible? The driver or the manufacturer? Worldwide, it is estimated that around a million people die in road accidents. Manufacturers and other advocates for driverless cars have used this fact to justify the huge investments required by their projects. No one ignores the risk that even a robot can fail when driving. Confirmation came very quickly, however. In March 2018, an Uber vehicle with a human behind the wheel but under the control of its autonomous systems hit and killed a pedestrian that was crossing the road in Tempe Arizona. Uber suspended its trials.

In *The Driver in the Driverless Car*, Wadhwa and Salkever (2017) discuss "the profound improvements in our lives that driverless cars will bring". Their diffusion, the two authors maintain, will reduce accidents, saving millions of lives, and reduce vehicle traffic in cities by a third to half. Most vehicles on the road are also looking for a parking spot, while self-driving cars used for car-sharing (which will become widespread in the meantime) do not need to park because they are continuously moving, picking up and transporting clients. In rush hour, these vehicles could be used 90% of the time. There will be fewer cars on the road as, collectively, self-driving cars could replace private cars.

As they do not have a steering wheel or other systems enabling human control, the vehicles will be lighter and thus consume less energy. An even more important advantage is that the cost of car-sharing with these vehicles will be a fraction of what it costs to keep a car. Owning a car for daily personal transport will no longer be worth it. This negative prediction is driving car manufacturers to enter the business of mobility services.

Self-driving cars will also provide benefits from a social perspective. People with disabilities will no longer have to struggle at length to find transportation, and women can find a car more safely even in the middle of the night.

It is commonly believed that the share of autonomous driving in the new car business is destined to rise continuously, despite various difficulties. Winkelhake (2017) reports on the conclusions of an IAO study, according to which many carmakers and suppliers predict that by 2030 fully-automated vehicles will be widely available. The prediction is based on the large number of projects, the size of the investments announced, and the declarations of intent of manufacturers and supplies. It is very probable that other companies—such as Uber, Baidu, Alibaba, and Tesla—could come onto the scene. Winkle also predicts that the fusion of services and autonomous mobility could give rise to new business models, whose offers will be easily accessible via internet-based platforms. If the legislative barriers and technological obstacles are removed, by 2030 autonomous vehicles could make up 15% of the market share in a middle scenario, somewhere between 4% in a conservative scenario, and 50% in a "high-disruption" one.

Waymo, which like Google is part of the Alphabet group, is considered to be at an advantage in the race to introduce a driverless car onto the market.[3] In October 2018, it was the first driverless car manufacturer to be authorised to go on the road in California without having an operator on-board ready to intervene.[4]

From the automotive industry, two types of response emerged: (1) a partnership between manufacturers or between them and investors; (2) a stand-alone strategy. Across Germany, Japan, and the U.S., partnerships are now common. Toyota struck up a partnership with SoftBank to create a range of services for self-driving vehicles. Honda has invested $2.7 billion into GM's Cruise[5]. One early leader, Waymo, has partnered up with Fiat Chrysler. Another tech company, Aurora, is working with Volkswagen, Hyundai, and the Chinese electric group Byton. Other carmakers have chosen to develop their own technologies. Toyota, Renault Nissan, Daimler, and Tesla are all working in-house.

Headwinds. "More consumers become skeptical about self-driving tech": this is the title of a study by J. D. Power in 2017, while automakers, tech companies, and suppliers invest heavily into self-driving technology. The results demonstrate that there is "significant trepidation" among consumers about autonomous tech, although the companies continue to insist on the advantages this could provide. Each age group, J. D. Power surveyed, except for those born between 1997 and 1994, "showed less trust in autonomous technology than they did in the survey of the previous year, even as they increasingly want semiautonomous safety technology".

Scepticism about the acceptance of driverless cars by consumers is very strong among American dealers. They are agreed in recognising that, "Big shifts are coming

---

[3] Waymo leads traditional automakers. "In the race to start the world's first driving business without human drivers, everyone is chasing Alphabet's Waymo. The Google sibling is ahead of traditional automakers such as General Motors, Mercedes-Benz and Audi by at least a year to introduce driverless cars to the wider public. A deal reached in January to buy thousands of additional Chrysler Pacifica minivans, which get kitted out with sensors that can see hundreds of yards in any direction, puts Waymo's lead into perspective." Source: "In self-driving car race, Waymo leads traditional automakers", *Automotive News*, 8 May 2018.

[4] Waymo the first without a backup driver. "Alphabet Inc.'s Waymo unit on October 2018 became the first company to receive a permit from California to test driverless vehicles without a backup driver in the front seat. Although self-driving vehicles are designed to obviate the need for a driver, most testing thus far has been with a safety driver behind the wheel who can take over in case of emergency. New regulations adopted by the state allowing companies to test on public roads without a driver with a special permit took effect in April 2018. The new requirements call for remote control technology, which allows for a remote operator to take control of a vehicle if the underlying autonomous system inside the car encounters problems." Source: "Waymo gets first California ok for driverless testing without backup driver", *Automotive News*, 30 October 2018.

[5] Honda Motor Co. spent roughly two years convincing Google's Waymo to share the technology of autonomous vehicles. In October 2018, however, it surprised the industry by taking another route, investing $2.75 billion with General Motors Co.'s self-driving unit. GM's Cruise unit and Waymo have been locked in close competition. Now, both are squaring off in a battle for leadership, having made splashes with two big-name partners. GM won backing from SoftBank Vision Fund a few months before securing Honda's investment, while Waymo has teamed up with Tata Motors Ltd.'s Jaguar. Source: "GM cuts in front of Waymo to seal self-driving deal with Honda". *Autonews News*, 4 October 2018.

to the auto retail world" and that they must be ready to face them. "But they won't turn the industry on its head, and they're not killing the personal-ownership model either." They consider levels 4 and 5 to be too complex and too costly to awaken the desire to buy.[6]

5 Electric Cars: Could They Create the Biggest Disruption Since the iPhone?

Traffic density is increasing and pollution from carbon dioxide emissions, particles, and noise is rising with it. "Over the past three decades, the dangers related to climate change have become increasingly clear and immediate", wrote Margo Oge M. in *Driving the future. Combating climate change with cleaner, smarter cars* (2015). For half a century, Oge continued, the promise of futuristic technologies has been made repeatedly in magazines and television shows, but in dealers' showrooms we have only seen modest technological improvements in new models. "Today, those promises are finally being realized." Driven by the digital transformation, the automotive industry has in fact sped up its response times with mobility services, lighter materials, less polluting engines, which are also lighter, but above all with electric vehicles (EV).

In the early days of car history, most cars on the road were electric-driven. In the U.S., at the beginning of the last century, 27 manufacturers offered electric cars. Among the customers leading the demand were society women. What appealed to women was that electric cars were simple and clean, with relatively few moving parts that could break. One advert said, "A delicate woman can practically live in her car and never tire". Yet the internal combustion engine (ICE) soon dominated the market. Two factors contributed to this (the same ones that should now bolster the demand for EVs): the higher range and a fast-growing petrol station infrastructure. By 1916, in the U.S. the market economics and the looming war had driven nearly all electric car companies out of business.

Fialka (2015) draws attention to the risks human beings run by not according importance to electromobility as a tool for defending the natural environment. "We ignore the new technological advances and science's warnings about climate change and drift along, retaining our old habits". He based his warning on the suggestions of a study by the University of California, which judged how disastrous the consequences

---

[6]"I'm not afraid of disruption" AutoNation CEO Mike Jackson said at a conference organised by Automotive News "Transitioning to autonomous vehicles that don't need a driver's supervision"—levels 4 and 5—"is exponentially complex and expensive," Jackson said. "The step up to Level 4 is the difference between putting a man on Mars and a man on the moon. We've put a man on the moon, but we haven't gotten anybody on Mars yet." The cost of a level 4 system could be between $100,000 and $200,000 per unit, he said. "No consumer is ever going to pay that kind of money for personal use of a Level 4 or Level 5 system." Source: "Keeping disruption in perspective", *Automotive News*, 5 October 2018.

would be if rapidly expanding economies, such as China and India, followed the example of the United States.

Tesla's case. In March 2016, the automotive industry responded to the question many manufacturers were asking themselves. Could the automotive industry still create the type of innovation that a company like Apple generates when it launches a new product? The response came from Tesla when the new Model 3, which, to anyone sceptical about the future of the electric car, said: price 35,000 dollars, therefore accessible to many, and 215 miles with one recharge, dissipating the concern about the limited autonomy of electric vehicles. In a few weeks, Tesla received 400,000 bookings from all over the world, accompanied by 1000 dollars in advance. Most analysts were cautious about the future of the initiative (from both the financial and industrial perspective), but they recognised that Musk and Tesla had brought about drastic innovation in the industry (Sumantran et al. 2017).

Is it real innovation? Wells and Nieuvenhuis (2015) compared Tesla's innovations with those introduced by other manufacturers previously. Smart (Mercedes) and Daewwo anticipated Tesla with a fixed price or "no-haggle" retailing. Smart represented a high-risk strategy for Mercedes in that it simultaneously introduced a new product destined for a new segment, a new factory with an innovative layout and process operations, a new brand, a new supply chain, new distribution and sales, and a focus on "green consumers". Mercedes did not make any profit with Smart. Many years of missed targets and losses only accumulated losses. Smart has been included among the cars that have accumulated the greatest losses in the history of the European automotive industry (see also Pellicelli 2014). The other innovator, Daewoo, also accumulated losses, but primarily due to an uncompetitive product. Daewoo was then taken over by General Motors.

Table 3 sums up the main innovations attributed to Tesla, and compares them with pre-existing samples.

A comparison. In the comparison between EVs and ICE-driven vehicles, it is necessary to consider the entire life cycle, production methods, services (in particular maintenance work) and, lastly, scrapping. A decisive factor that supports the diffusion of EVs is energy recovery. Electric-powered vehicles have a series of clear advantages: (1) High energy effectiveness. In an EV, the effectiveness is 95%, while in an ICE engine, it is around 35%; (2) Assembly is simpler because in EVs there is no gearbox, fuel system, or exhaust system; (3) "Engine brake" effect. While with an ICE, energy is lost during breaking, in the EV it is recovered and accumulated ("fed into the battery"); (4) smooth power delivery and quiet operation; (5) reduced running costs chiefly due to the lower cost of using electricity (not entirely compensated by the initial higher purchase cost); (6) after disassembling GM's Chevrolet

Table 3 A comparison of the business model innovations (and marketing practices) from Tesla with pre-existing innovations

Tesla innovation	Pre-existing examples
Ownership of retail outlets	Very common and long-established practice in Continental Europe: adopted by Daewoo as a market entry strategy in the UK in the 1990s
Creation of "boutique" retail or experience outlets in shopping malls and other mixed retail locations	Attempted by Smart during the early phase of marketing. Parallel examples include the Toyota Amlux in Tokyo and the VW Volfsburg "Autostadt" brand experience facility
Fixed price, "no-haggle" retailing	Adopted by Daewoo as a market entry strategy in the UK in the 1990s
Provision of free access to unlimited charging via own fast-charger infrastructure (30-min recharge)	No comparable example. However, many instances where new cars were offered with a 12-month supply of petrol
Car built to order, not "sold off the lot"	Very common in Europe, particularly for prestige and sports cars, for at least a proportion of the total output. Morgan is a good example
Ordering new cars via retail outlets or the internet	Internet retailing is well-established, though because of legal constrains orders still need to be rooted via dealerships
Introducing new brand, new model and new manufacturing facility simultaneously	An even more ambitious version of this was attempted at launch by Smart with the Hambach plant

Source Beeton and Meyer (2015, p. 7)

Bolt, UBS Group concluded that it required almost no maintenance, with the electric motor having just three moving parts compared to 133 in a four-cylinder internal combustion engine.

The EV should not be an evolution of the traditional car. "The EV must in a definite manner be designed and born electric. This will fundamentally change the vehicle typology used today in common combustion engine vehicle or hybrid electric vehicle."

Heavy investments planned. According to the consulting firm AlixPartners Globally, carmakers and suppliers will invest an estimated $255 billion into electric vehicles up to 2022, compared with about $25 billion in the previous eight years. Electric-driven cars now make up an extremely modest share of the European fleet. In 2017, there were a total of 826,000 battery-powered and plug-in hybrid cars on European roads, according to AlixPartners. This corresponds to the share of around 0.32% of Europe's 259.7 million vehicle car fleet.

Various factors slow down the expansion of electrified vehicles (Coffman et al. 2017). Compared to equivalent conventional cars, EVs have a high initial purchase price, the cost of the battery makes up a significant proportion of the price of a vehicle, they are more sensitive to weather and other driving conditions, have uncertain rates of depreciations, lower expectations with regard to the longevity of the powertrain, there is a low availability of public recharging infrastructure, especially outside of large cities, and there are concerns over the range and the time required for recharging (Beeton and Meyer 2015). The "energy density" of the batteries nowadays is much lower than that of diesel and petrol. This influences the weight and dimensions of the batteries, and therefore the relatively shorter autonomy. The disadvantages concerning the battery and the longevity of the powertrain can be reduced in the short to medium term thanks to technological advances.

Forecasts as to when this will be possible are difficult to make. Some of the latest battery technologies became obsolete before reaching the market as a result of the speed of progress in the industry. The solution to problems concerning access to an external recharging network depend mainly on public investments (Harrison and Thiel 2017). Some manufacturers, including Tesla and Volkswagen, are intervening directly with their own recharging networks. Public intervention is expected for other aspects, too. With the objective of encouraging research to improve battery efficiency, the EU is allowing state aid for electric battery research and will offer billions of euros of co-funding to companies willing to build giant battery factories.

China is by far the leading market for electromobility by volume and growth. Growth is driven by state subsidies and by the large network of recharging facilities. The result is that almost 40% of the EVs on the road worldwide are on Chinese roads, with over 90 models.[7]

Despite the uncertainties, there are many projects directed at launching new EV models. By 2020, 60 battery-electric and plug-in hybrid models should join the market that already includes Tesla, the Nissan Leaf, and the Chevrolet Bolt.[8]

Some carmakers introduce EV models into their product portfolio to reduce the average CO_2 of their fleet. For example, Bentley has launched its first full-electric vehicle to help reduce its fleet's CO_2 emissions as lawmakers throughout the world introduce stricter reduction targets. Bentley has one of the highest carbon footprints of any brand in the group due to its reliance on heavy sedans and the large Bentayga SUV.

---

[7]"The formula for boosting vehicle electrification is not complicated: combine hefty purchasing subsidies with a vast recharging network and sales will take off almost immediately. China did this, not only to clean the air in some of the most polluted cities in the world but also to accelerate the development of a domestic auto industry that was lagging behind Western competitors in terms of internal combustion engine technology. The result? Almost 40% of the 3.2 million full-electric vehicles worldwide are in China, where consumers can choose from 92 models. They also can use one of the more than 241,000 charging stations—more than half of the 424,000 available worldwide, based on data from turnaround specialist AlixPartners." Source: "Why China is by far the world's largest market for full-electric cars", *Automotive News*, 5 August 2018.

[8]"... make up a tiny fraction of U.S. light-vehicle sales, but plenty of competition is on the horizon, with more than 60 battery-electric and plug-in hybrid models slated to reach dealerships by the end of 2020". Source: "Nearly 100 electrified models slated to arrive through 2020", *Automotive News*, 5 October 2018.

A non-existent market. On the eve of the 2020s, a real market (in terms of size) of EVs does not exist. Various attempts have been made in the past to introduce EVs, but without any significant success in terms of volume. Besides the comparison of pros and cons mentioned earlier, the main obstacle is that it has never been possible to reach a sufficient sales volume to trigger economies of scale. If a market does not exist, it cannot be analysed. Market research cannot give any indications. A prospective customer can only say what their buying intentions and estimations are if they know a product and have used it before, even in configurations not exactly the same as those proposed by the trailblazer.

As Beeton and Meyer (2015) observed, the battery seems to be a source of paradox. When batteries are produced in higher volumes, their cost will go down, which could make EVs competitive with current hybrid and ICE vehicles. However, to sell EVs in higher volumes, the cost of the battery needs to be low. Moreover, to beat the reluctance of those that fear that EVs are unreliable, they need to become more widely used. Competitions between electric-driven cars (the current FIA FormulaE) can contribute to this, as they did a century ago for ICE-driven cars. Perhaps a world championship could push EVs towards the tipping point. State intervention could certainly be more effective with subsidies for buyers and the construction of recharging networks (as mentioned above).

Waiting strategies. Instead of investing in models that will not generate profits, some incumbents prefer to invest most of their resources in systems for the mass production of new models starting from 2030. They anticipate that by that date, batteries will cost noticeably less than they currently do, and will have longer autonomy. Consequently, the rising demand for EVs should make it possible to achieve economies of scale.

Experts interpret this choice as giving up on confronting Tesla in the short term, preferring to be ready when the competition to dominate the market truly begins. They are convinced that competition will occur in factories and not in car showrooms. They rely on their greater experience in high-volume production and on the prediction that the demand for EVs will long remain a small proportion of the total demand for cars worldwide.

Wall Street investors hold the opposite opinion. They anticipate that other competitors will enter the automotive industry given the weakening of the barriers to entry. Mainly for this reason, market capitalisation by incumbents has fallen to an all-time low, while Tesla, despite production and governance problems, maintains a very high capitalisation compared to that of GM.

As always occurs in the face of uncertainty, alliances present a way to reduce risks. Many carmakers are jointly developing electric vehicles (as they are doing for self-driving cars) in strategic alliances meant to save the companies billions of dollars. For example, Volvo and the Chinese internet giant Baidu will jointly develop electric vehicles capable of level-4 autonomous driving. This is another move by Volvo to become a leading player in the robotaxi market. Volvo Cars has acquired a stake in a San Francisco-based vehicle charging company called FreeWire Technologies, as part of its plan to electrify its lineup.

Some manufacturers are examining the possibility of totally giving up on ICE for some brands, and only adopting EVs for these. In 2018, under a strategy being considered by company leaders, the British luxury marque Jaguar announced that it would replace its conventional vehicles with battery-electric models over the next five to seven years.

A chorus of dissenting voices. Keith Crain, Editor-in-Chief of *Automotive News*, asks whether manufacturers that, around the globe, are investing billions, had asked their customers, or potential customers, whether they were interested in purchasing a battery electric vehicle. There is no shortage of companies developing electric vehicles. "It seems that everyone has committed to it, and no one knows for certain whether there is a market." Confronted with uncertainty, Crain appreciates the prudent strategy by GM, which seems to have the right approach, having introduced onto the market both a plug-in and a hybrid at the bottom of the range, rather than designing and developing costly vehicles without any certainties as to the volumes of demand.

Other experts in the industry and carmaker manager express caution, or even skepticism. Fearing that drivers are not ready for this new technology and, while there are no certain elements for predicting the times and intensity of the demand with a reasonable degree of certainty, carmakers are still planning a tidal wave of battery-powered models. The head of product development and purchasing of Ford said of electric cars: "Nobody can cite what the actual demand will be."

Another expert in the industry presents a more drastic opinion: "Automakers with ambitious plans to roll out more than a hundred new battery-powered models in the next five years appear to be forgetting one little thing: Drivers aren't yet buzzed about the new technology."

However, the major players are investing, fully aware that the road ahead will be a long and uncertain one. Volkswagen AG is preparing for a steep learning curve in the transition to autonomous, electric and connected vehicles. Volkswagen has planned to introduce the first vehicle in its new electric and connected vehicle product line in 2020. Yet Burkhard Huhnke, Volkswagen's senior vice president of e-mobility, said the automaker expects it will take years to bring new vehicle technology "nearly to perfection."

We should not forget, moreover, that the electric car needs to be totally reinvented. It has no steering column, brake pedals, accelerator pedals, or emergency brake pedal. It does not have a gearshift panel. To achieve the greatest efficiency, it needs to be designed "born digital". It cannot be an evolution of a traditional car or current hybrid models. For established manufacturers, there is also the problem of production plants: when should they be built? In the long term, an increasing acceptance of EVs should present itself, in line with the possibilities of greater driving comfort and more attractive prices. However, demand will remain low for years. One solution, to hold off for the timebeing, adopted by Mercedes Benz involves producing EVs along the same lines as models with combustion engines, "enabling us to respond flexibly to demand and use plant capacity to best effects", as the manager of the car production and supply chain of the German group declared.

A viable explanation. Why, even though it is difficult to predict the consistency of EV demand, are almost all manufacturers investing significant amounts? One response to the doubts of Crain and other commentators and experts can be found in strategy tests. Sometimes, when several competitors are faced with uncertainty as to the outcome of an investment, and are wondering whether or not to compete, it can be better for them to all risk all losing rather than not take part in the contest and see one or a few be successful or gain advantages that are difficult to bridge.

In the 1970s, drilling work in some points of the North Sea revealed the presence of oil deposits. Each nation bordering on the North Sea was assigned a share of the surface to explore. When Great Britain put its drilling and extraction rights up for sale, there was no response by the oil companies for a certain amount of time. There was no certainty as to the submarine extension and the significance of the deposits. Everyone was afraid of the risk of not finding oil or of not finding a sufficient amount to justify the investment. When, after the delays, British Petroleum acquired the rights to certain parts, it was followed first by French Total, and then by all the major companies. The reasoning was simple: if we do not follow and there is enough oil to make the investment profitable, the competitor that is successful will gain a highly advantageous position for transport costs for the supply of oil in Europe (compared to supplies from the Middle East or Venezuala). If there is no oil, or there is not a profitable amount, we all lose, but the initial competitive positions remain the same (Pellicelli 2014). To the question "How does PSA Group justify big investments in mobility when the revenues are a fraction of making, selling and servicing cars?", the head of PSA Group's new mobility unit responded, "This is a strategic choice. If autonomous cars arrive, we will need to be in the mobility market to avoid losing our relationship with the customer. We want to provide solutions that customers will need."[9]

References

Beeton D, Meyer G (eds) (2015) Electric vehicle business model. Global perspective. Springer
Coffman M, Bernstein P, Wee S (2017) Electric vehicles revisited: a review of factors that affect adoption. Transport Reviews 37 (1):79–93
Fialka J (2015) Car wars. The rise, the fall, and the resurgence of the electric car. Thomas Dunne Books, US
Firnkorn J, Müller M (2012) Selling mobility instead of cars: new business strategies of automakers and the impact on private vehicle holding. Bus Strategy Environ 21(4):264–280
Foster R, Kaplan S (2011) Creative destruction. Crown Business

---

[9]"PSA exec outlines risks and rewards from leap into mobility services", Tiny titans, *Automotive News Europe*, September 2017.

Harrison G, Thiel C (2017) An exploratory policy analysis of electric vehicle sales competition and sensitivity to infrastructure in Europe. Technol Forecast Soc Chang 114:165–178

Jansson J, Andervin M (2018) Leading digital transformation. DigJourney Publishing

Jullien B, Pardi T (2013) Structuring new automotive industries, restructuring old automotive industries and the new geopolitics of the global automotive sector. Int J Automot Technol Manage 13(2):96–113

Moore G (1965) Moore's law. Electronics Magazine 38(8):114–117

Oge M (2015) Driving the future. Combating climate change with cleaner, smarter cars. Arcade Publishing

Pellicelli G (2014) Strategie d'impresa. Egea, Milan

Ryan D (2015) Understanding digital marketing. Marketing strategies for engaging the digital generation. Kogan Page

Simoudis E (2017) The big data opportunity in our driverless future. Corporate Innovators

Sumantran W, Fine C, Gonsalvez D (2017) Faster, smarter, greener. The future of the car and urban mobility. The MIT Press

Venkatraman V (2017) The digital matrix. New rules for business transformation through technology. LifeThree Media

Wadhwa V, Salkever A (2017) The driver in the driverless car. Berrett-Koehler Publisher

Wells P, Nieuvenhuis P (2015) EV models in a wider context: balancing change and continuity in the automotive industry. In: Beeton D, Meyer G (eds) Electric vehicle business model. Global perspective. Springer

Winkelhake U (2017) The digital transformation of the automotive industry. Springer

Chapter 19
Towards the 2030s: Unusual Times Call for Unusual Strategies

Abstract Three main issues have to be analyzed when defining a marketing strategy in the time of digital transformation. The first one is the consumers' desires, which are increasing to such an extent that they might be described as "unreasonable expectations". If consumers want a quote for a built-to-order car, they expect fast responses even for the most personalised, specific requests. On-board instruments have been improved: voice-activated functions, artificial intelligence, and augmented reality. The problem lies in understanding, during the design stage, which technology actually brings value for the client. The second issue to analyze is the competitive environment. The digital revolution redefines the competition between companies and the relationships between companies in various ways, and competition occurs less within individual industries and more between different industries (Google, Apple, Amazon, Uber, Lyft, Didi). Moreover, some companies compete among themselves in certain areas, but are partners in others (co-opetition). The third issue deals with understanding how digital marketing fits into the company's business model and how and what changes are needed in the processes. Big data must be managed; the nature of competitive advantages can change; new value propositions have to be offered; and the need of new organizational structures and culture increases. Under the pressure of the digital transformation, established carmakers must make broad adjustments to marketing strategies. They must defend "old" business segments and open new ones. The key questions are: How is selling mobility services different to selling vehicles? and "A new machine will change the world again"?

Defining a new marketing strategy. In their publications on digital marketing, Ryan (2015), Dodson (2016), and Chaffey and Chadwick (2016) agreed upon the identification of three key phases to define a marketing strategy in the time of digital transformation. Given that there are no foolproof formulas for all situations, since each company has its own set of circumstances to tackle and must find unique solutions, the three phases are as follow:

(1) Knowing your prospective customers. Who are they? What do they expect? With new offers, is our marketing directed at the "old" clients offline or at clients in the new demographic segments? How can we reach them? Who are they and want do they want from us? With digital technology, are we reaching the same clients as before or a new demographic in new segments? How does our target use digital technologies? Unlike the traditional communication channels, looking

E. Candelo, *Marketing Innovations in the Automotive Industry*, International Series in Advanced Management Studies, https://doi.org/10.1007/978-3-030-15999-3_19

at what people do online allows companies to understand a lot about its clients. *Know your customers.*

(2) Know your competitors. Who are they? Are they the same as in the offline phase? How can we differentiate our online offer from theirs? It is important to remember that new competitors often emerge in unexpected ways and at unexpected times. Just as the digital age offers us the chance to reach geographically distant markets, new competitors from far-away places can also reach our market. *Know the competition.*

(3) Understand how digital marketing fits into our own business model. How and what changes in our processes? Is our product suited to online promotion? *Know your business.*

What is more, the rules for any long-term plan also apply. First of all: where do we want to go? What results do we want to obtain from a marketing plan based on digital technologies? Increasing online sales? Increasing the awareness of our brand? Secondly, but no less importantly, it is necessary, taking advantage of the characteristics of digital technology, to constantly compare the results obtained to the expected ones (this is particularly true for the results of advertising investments).

Let us examine the three phases with the most important details.

1 Know Your Customers: The Consumer Context is Changing

The dominant model of the mechanical age and the first phases of the electronic/software age, as discussed in the previous chapters, were based on the creation of efficiency through economies of scale in production (the higher the production volume, the lower the unit cost per product) and economies of scope in distribution (the higher the number of clients, the lower the distribution cost of each unit produced).

In the lead-up to the 2020s, mass markets have made increasing room for fragmentation in various segments and niches. What is more, many markets, segments and niches are replaced with customer networks in which the actors exchange information and evaluations among themselves, influencing each other's choices reciprocally.

The relationships between consumers and the company offering them products or services also change. Thanks to the advances in technology, traditional "one-way" communication—inform, persuade, and convince to buy—AIDA (Awareness, Interest, Desire, Action), and "buy or do not buy"—has been replaced with "two-way" communication in which the company is still the main actor, but the consumer acquires greater power, to the point of sometimes becoming the main influencer (still at a moderate level for cars, given the complexity and specificity of the product).

Technology is changing consumer behaviour. Buyers are better informed, communicate more among themselves, interact, create virtual communities, and have more access to information than in the past. Digital technology allows for faster connections between consumers, overcoming geographic or political boundaries. Thanks to the new technologies, consumers can carry out research, and compare products and services. Technology partly shifts the power of information in favour of the consumer, taking it away from the producer. Consumers are more involved in the production of the product of service they intend to purchase. They have a better understanding of what they can ask and can better specify it to the producer (the rise of the "prosumer"). Mass production and mass marketing are partly over.

If the customer network gains weight, the company must understand the interactions established between customers and prospective customers, understand how opinions are formed and their responses, and how they can both influence the reputation of the brand. Traditional marketing tools, from printing to TV and direct marketing, can retain their capacity to inform and convince, but they need to be reconsidered because their cost effectiveness can decrease or even vanish. This is especially true when buyers belong to new generations.

"On-demand, any time, any place, anywhere". Technology makes it possible to speed up the buying process and buying times for a product. Consumers can meet their expectations more quickly and more easily compared to the recent past. In the digital economy, geographical distances and times can no longer pose barriers.

Consequently, the expectations of consumers from the new generations have increased greatly. They have at times been described as "unreasonable expectations". These consumers, accustomed to automatic responses, expect the same speed at all times and from whoever offers products or services. If they want a quote for a built-to-order car, they expect fast responses even for the most personalised, specific requests.

In order to respond to growing expectations and at the same time create differentiation, car manufacturers, also driven by competition, have modified their offer, adding services to each product. However, responding to clients is becoming increasingly challenging, especially if waves of innovation intervene. On-board instruments have been improved: voice-activated functions, artificial intelligence, and augmented reality. The problem lies in understanding, during the design stage, which technology brings advantages (value for the client) that the client can appreciate in concrete terms. Rogers (2016) has summed up the main changes from the Analog Age to the Digital Age in the following way: (a) from the customer as a member of a mass market to the customer as a member of a network; (b) from one-way communication to two-way communication; (c) from the company as the key persuader to a network of customers as key influencers; and (d) from marketing as a tool to persuade to marketing as a tool to inspire buying decisions and building loyalty.

A mass upheaval in the automotive market. According to Winkelhake (2017), the car industry is the perfect example of the new environment created by digital transformation. Towards the 2030s, two trends are generating a profound change in the automotive market: (1) new technologies shake up the existing context, and new competitors threaten to enter the market; and (2) rapid evolution in consumer behaviour in transport demand. Below, we examine the effects indicated in point (2), postponing the examination of point (1) until further on, concerning the competitive context.

Table 1 summarises how various trends are changing the car industry in advanced markets (Winkelhake 2017).

"Multigraphic". Unlike in the past, nowadays people adopt a lifestyle with shorter durations and more varied forms. For manufacturers, the consequence of this is the need for sharp, more precise segmentation, and a personalised offer of mobility services, like Audi's "Select" and "Shared Fleet" offer.[1]

Table 1 Consumer trends in the automotive market

Consumer trends	Implications for (vehicular) mobility
Multigraphic	More fragmented life-design needs are becoming more situational. "Stage-of Life-Products" are becoming more important than target group strategies (age, social status, etc.)
Downaging	Consumers are feeling younger than their biological age. No "ghetto-products" but they experience products through a second awakening
Family 2.0	Network, patchwork and fragmented families have a higher and highly differentiated need for mobility, which can be catered for solely by a family SUV or station wagon
Neo-cities	Vehicular mobility, which adjusts to the requirements of future green cities (zero-emission-cities)
Greenomics	Vehicular mobility which at the same time satisfies a healthy, sensual life style Mobility solutions, which are ecologically friendly, but also sustainable for the consumer
New luxury	Products which increase one's quality of life. Nevertheless, there is a trend away from prestige and status objects
Simplify	Simplification, time saving, simplicity, invisibility of technological processes
Deep support	Support services which cater for the individual's needs. Infrastructure of micro services which organise life between home and work
Cheap chic	Affordable, "clever" products, which nevertheless satisfy the desire for exclusivity, design, and luxury

Source Winkelhake (2017, p. 80)

---

[1] **"Select".** The client does not acquire possession of a specific vehicle, but rather the right to use, one after the other, up to three types of second-hand but recently built vehicles: a convertible for the summer, a spacious car for specific needs, and a sedan for long trips. **"Shared Fleet".** It is offered to companies that have vehicle fleets for employees. Through a booking portal, employees can use the company car for private trips outside of working hours. Vehicle usage thereby increases and the company offers employees a mobility alternative.

Downaging. People who are "no longer young" are physically better off than their contemporaries of previous generations, and their share of the total population increases. If they have economic means, they are willing to pay for a car that is easy to get into, a suitable steering wheel and seats, and electronic driving assistance.

Neo-cities. On the one hand, this trend reflects the increasing urbanisation (more people live in large cities), and on the other hand the commitment of the administrators of these cities to making them "greener" and leading them towards "zero-emissions". There are plenty of examples of traffic restrictions. London was a pioneer in this regard. San Paolo allows cars to drive on alternate days (cars are identified by their number plate). During rush hour, Singapore only allows vehicles with at least three people on-board to be on the road. Evidently, every traffic restriction has a negative impact on sales volumes in the car industry because it forces people to choose alternative means of transport.

Other trends also emerge: *New Luxury*. This is a trend concerning people's behaviour. They move increasingly towards immaterial values, such as quality of life, rather than towards seeing the car as a status symbol. *Simplify* reveals the growing number of people that expect only essential technology from a car. *Cheap chic*. This is a trend towards quality and premium, at a reasonable price.

2 Know the Competition: The Competitive Environment is Changing

As a result of the digital transformation (Jansson and Andervin 2018), incumbents not only need to confront and respond in a new way to rivals in the automotive industry, but must also confront new rivals from outside of the confines of the automotive industry and which take clients from them.

In a previous chapter, we noted how, in the car industry, the increasingly widespread use of platforms has changed the competition and relationships both between companies and between companies and their clients. Up to the early 2000s, competition occurred in a context with stable borders, between similar companies (substantially assemblers), albeit with different characteristics: companies with strategies based on large volumes; those with premium strategies; and those with niche strategies. These companies created value with similar business models based on a capital-intensive approach to purchasing plants and investing in R&D, a search for efficiency based on economies of scale (and scope), the assembly of products from suppliers, and the delivery of the final product to the consumer through a network of dealers.

As summarised in Table 2, the digital revolution redefines the competition between companies and the relationships between companies in various ways, and competition occurs less within individual industries and more between different industries. Lastly, some companies compete among themselves in certain areas, but are partners in others (co-opetition).

Table 2 Porter's five forces model and internet effects

Bargaining power of buyers	The power of online buyers increases as they have a wider choice and prices are likely to be forced down through increased customer knowledge and price transparency
Bargaining power of suppliers	When an organisation purchases, the bargaining power of its suppliers is reduced since there is a wider choice and increased commodisation due to e-procurement and e-marketplaces
Threat of substitute product and services	Substitution is a significant threat since new digital products or extended products can be readily introduced
Barriers to entry	Barriers to entry are reduced, enabling new competitors, particularly for retailers or service organisations that have traditionally required a high-street presence or a mobile salesforce
Rivalry between existing competitors	The internet encourages commoditisation, which makes it less easy to differentiate products. Rivalry becomes more intense as product lifecycles decrease and lead times for new product development decrease

Source Analysis and synthesis of documentation taken form Porter (2001), Porter and Heppelmann (2014), and Chaffey and Chadwick (2016)

Rogers (2016) has identified a set of competition changes in strategic assumption from the Analog Age to the Digital Age in the following way: (1) from competition within well-defined boundaries to competition across porous boundaries of industries; (2) from a well-defined distinction between partners and rivals to an unclear distinction between partners and rivals; (3) from competition as a zero-sum game to cooperation among competitors in key areas; (4) from strategic assets held within a company to strategic assets distributed across an external network; (5) from uniqueness in product features and benefits to a platform with partners; and (6) from a few dominant companies in each industry to a winner that takes everything, building on network effects.

The value chain and Porter's five forces model, which became common practice in analyses of the economic sector, were based on well-defined borders in the automotive industry. Nowadays, digital technologies are dismantling these borders or rendering them more permeable.

In 2001, Porter revised its five forces model to take into account the impact of the internet (Porter 2001). The main effects are summarised in Table 2.[2]

---

[2] According to Porter, the most significant effect of the internet concerns buyers' bargaining power. Increased buyer power is important both for B2Cs and for B2Bs. B2Cs have a greater capacity to compare prices and assess products before buying. For B2Bs, the effect is even more pronounced, both in terms of the force of direct contacts and online auctions. The overall effect is driving down prices, especially for high volume-low price products.

Tesla competes in passenger transportation with Daimler, BMW, and Lexus (Toyota), but has a business model that also includes the battery production sector. Hence, someone asks: "Is Tesla a car company or an electric battery car company? We don't know yet." For Venkatraman (2017), Tesla is a car company, in the conventional sense, because it designs and develops cars, just like GM and Ford, but it is not a car company because it provides electrical charges in home garages and in a network of supercharging stations (while traditional carmakers do not provide free gasoline).

Apple and Google (which are attracted by the estimated 25 gb of valuable data a connected car can generate every hour) pose a serious threat to established firms in the automotive industry by introducing the autonomous vehicle. They are particularly threatening as a result of their high degree of liquidity.

Uber, Lyft, Didi and others base their growth on network effects. The value of platforms increases with their increased use by customers. Instead of investing in the acquisition of a fleet of cars and renting them out to anyone requiring transportation, these software-powered competitors provide an online platform that allows car owners to rent out their car when they are not using it to travellers that need it. Uber and others offer more value to consumers than traditional car makers and rental companies: better price and a more widely distributed offer across the territory (due to the number of drivers). Incumbents struggle to respond with the same business model, and their position on the market is therefore weakened.[3]

New challenges for marketing. With the change in the competitive environment in the automotive industry, marketing must face new challenges. The main challenges and their causes are summarised below.

- Digital technologies weaken barriers, meaning that new competitors can enter the industry more easily than in the past (i.e., Uber and others in the automotive industry). Many small companies also enter, which often become startups, thereby increasing the rate of innovation.
- Digital technologies take power away from organisations that offer products or services, while they increase that of those buying.

---

[3] Rogers (2016) distinguishes between symmetric competition and asymmetric competition. Symmetric competitors offer clients similar value propositions. For example, BMW, Daimler, and other manufacturers have different brands, address different drivers, but their private vehicle ownership or leasing offers have similar characteristics. Symmetric competitors create value for the client with the same business model. They can be volume or premium manufacturers, small- or large-sized, but they all have supply chains, production plants, and dealership networks.

Asymmetric competitors are very different. They offer similar value propositions to clients (i.e., mobility), but their business models are different. For a manufacturer like BMW, Rogers observes, an asymmetric competitor could be a ride-sharing service like Uber, because customers buy fewer cars since Uber satisfies their transport requirements.

- The weakened barriers and consequent easier entry for newcomers make it difficult to identify the threats of new competitors. Often, they come from other industries, unexpected, "invisible until it is too late". One effect of digital transformation is that competition comes from outside the traditional boundaries of the industry. Perkin and Abraham (2017) call this "horizontal innovation": Bosch is developing digital services for urban mobility, thereby entering into competition with some of its automaker clients. In 2018, it introduced an electric van-sharing service in Germany, expanding in the field of pay-per-minute vehicle rental services. Bosch estimates that an electric van-sharing service has a high potential for growth (+).
- History shows that during profound changes in technologies (as is occurring now with the digital transformation), attackers, and not incumbents, have the economic advantage. Although at the outset they have small production volumes and still cannot rely on large economies of scale, attackers often cause the decline of companies that are "long-term survivors" and of established marketing strategies. The evolution of marketing strategies in the automotive industry confirms this assertion. "In the long run, markets always win" (Foster 1986; Foster and Kaplan 2011).

To attempt to predict where marketing strategies are headed in the automotive industry, it is important to remember that we could be faced with a radical innovation, rather than an incremental one as occurred for a long time in the past (if EVs, autonomous driving and ride-hailing reached "near perfection" at the same time, the situation would change radically).

As Foster and Kaplan have recalled, in the automotive industry change has always been slow compared to that of other industries. Innovation in the production process has been incremental in nature, without any major jolts. For 70 years, manufacturers transformed raw materials into final products, controlled costs carefully, and sought and achieved large economies of scale. Their production process has remained essentially unchanged, as has the product configuration. In terms of structure and functions, production plants have remained the same. Products have been sold through the same distribution channels. Incremental innovation did not call the business model into question.

Bosch wants to avoid competing against traditional automakers in areas such as vehicle manufacturing, lest this should alienate clients that buy its components and systems. With new rivals such as software companies, both auto suppliers and automakers seek to acquire positions in smartphone-based mobility as a preliminary action to manage fleets of robotaxis. "Self-driving cars would enable automakers and software companies to enter the ride-hailing business without having to pay drivers. Bosch is already collaborating with Mercedes-Benz to develop self-driving cars". Source: "Bosch will launch electric van-sharing service to compete with automakers", *Automotive News*, 9 October 2018.

Tesla's competitors are coming. Tesla dominated the plug-in electric vehicle market for a long time, with little direct competition in the market from $60,000 upward. This was also thanks to generous tax subsidies. Until the Chevrolet Bolt came along, with more than 200 miles of range on a charge, most other electric cars barely reached 100 miles of autonomy. Those days are over.

On the eve of 2020, the competition for electric car buyers has become more intense, with several well-engineered luxury vehicles from well-funded European manufacturers. Porsche, Audi, Mercedes, and Jaguar have all launched vehicles that will create real competition for Tesla. Moreover, this is happening as federal tax incentives begin to wind down for Tesla, which has reached the 200,000-sale threshold at which federal EV credits start to phase out. The wave of premium European carmakers includes Porsche with the Taycan, Audi with the e-tron, and Mercedes-Benz with its EQ range. In 2018, Jaguar started to deliver its I-Pace, a full-electric model with comparable entry-level prices to Tesla's Model S sedan and Model X SUV. The Jaguar I-Pace comes with a descent of luxury, racing, and engineering expertise.

While the premium European carmakers attack Tesla in the higher price segment of EVs (Tesla's Model S), Volkswagen occupies the lower price segment (Tesla's Model 3). Volkswagen has a plan to put 50 models on the road across the group, including volume brands such as VW, Seat, and Skoda. The world's largest automaker plans to use its massive scale to offer less expensive EVs than rivals like Tesla. In particular, the German group of brands plans to add a subcompact crossover costing about €18,000 ($21,000) to its all-electric I.D. family of vehicles.

3   Know Your Business: The Company Context is Changing

As a consequence of the above-mentioned trends, the company context is changing profoundly. Among the various aspects of the change driven by the digital transformation, some stand out in terms of importance: (1) With the continuous increase in the number of connected devices and services, the data they generate increase exponentially (Big Data); (2) the nature of competitive advantages changes; (3) there is an increased need to constantly adapt the value proposition; and (4) the need for new organisational structures and culture increases.

1) Big Data
Even data have acquired a new value. Nowadays, the automotive industry considers data to be an asset to be used better than in the past. It seeks greater efficiency and more rapidity in drawing useful information from data.

In the past, data were generated within the company, bearing costs to understand, measure, and assess. The data were used to analyse, compare, manage, and predict. The most important data included: (a) the production and distribution costs, useful for providing a basis for prices, assessing efficiency, and evaluating projects to launch new products; (b) data on the development of stocks and sales; and (c) data on the cost-benefit ratio of knowledge acquired. While these data were focused on the inside of the organisation, others were aimed at evaluating the impact of trends in the external environment (economic, social, legal, and technological) on the company.

Nowadays, technologies make it easier to collect large quantities of data, which make reporting easier and reduce the risks of errors. There are data on the behaviour and expectations of clients and prospective customers, data on visibility tests for advertising, data to understand which offers respond best to the expectations of a certain target of a particular audience, and data on survey responses. Even customer relationship management tools supply large quantities of data.

An even larger quantity of data is generated outside the company (i.e., Bloomberg for economics, demographics, etc.). Their acquisition is relatively simple. One problem is how to extract useful information from the large amount of data available. Competitive advantages can be drawn from extracting trends and identifying possible targets (McGrath 2013).

While the data collected and processed within the company have an objective (which data and for what purpose) and a system of procedures, and hence a structure, those drawn from outside do not. Tools are therefore required to analyse and extract what management needs (Leal-Millán et al. 2016). Advances in Artificial Intelligence (AI) make it possible to extract patterns and trends more easily than in the past.

Big data exploitation using machine learning. According to Simoudis (2017), machine intelligence (an expression the author prefers to machine learning) is key for the exploitation of big data in autonomous vehicles and the next generation of mobility solutions. Machine intelligence is part of Artificial Intelligence. Concerning the programmes and methods that allow computers to acquire new capacities without being explicitly programmed, "but only by being exposed to new data", Google, Tesla and Uber have for some time "been collecting live data from the sensor of the vehicle they operate".

Machine intelligence can be used for data application which, according to Simoudis, can: (1) foresee "the behaviour of other vehicles" in the vicinity of a driverless vehicle, as well as of a pedestrian; (2) produce a "complete understanding" of the vehicle and its passengers; (3) increase the quality of the "vehicle's usage economics", which is especially useful for fleet management companies; and (4) provide a "personal transportation experience" to passengers and drivers.

Tesla is disrupting the use of big data. For example, the telemetry each vehicle is equipped with is used to analyse the entire fleet's usage patterns, discover crashes, and identify maintenance needs. Through the mass of data collected by Autopilot, Tesla is creating its own high-definition map needed for autonomous driving. At the core, Uber is also a big data company. In addition to having acquired Microsoft's mapping assets and knowhow, it is making a large investment in developing high-definition maps.

2) Change the nature of competitive advantages

Another effect of digital transformation is the change in nature of competitive advantages (Porter and Heppelmann 2014). In *The End of Competitive Advantage*, Rita

Gunther McGrath (2013) argues that it is time to go beyond the very concept of sustainable advantage. McGrath based her conclusions on research that examined companies with a market capitalisation of over \$1 billion and which, in the period 2000–2009, achieved a net income growth above 5% compared to the global rate of GDP growth. Only ten companies passed this test.

Examining their strategies in depth, McGrath drew useful lessons for companies that are now facing the digital transformation to avoid "the risk of being trapped in an un competitive business". These are: (a) continuously reconfigure the product and strategies (achieving a balance between stability and agility); (b) do not defend a declining competitive advantage indefinitely (healthy disengagement); (c) direct resources at ensuring agility and flexibility for the organisation (using resources at the location to promote deftness); and (d) move from episodic innovation to continuous, systematic innovation (leadership and mindset of companies facing transient advantages).

The Boston Consulting Group (BCG) arrives at similar considerations and suggestions. BCG takes up the four-quadrant matrix from the 1960s again, and reconstructs it taking the digital transformation into account, in order to study its effects. They concluded that companies must constantly reconsider their competitive advantages, speeding up the time needed to move resources from one business to the other (if many businesses, as in the classic model), and therefore from one quadrant of the matrix to the other. The research (which compared a five-year period from 2008 to 2012), revealed that over the years, businesses moved between quadrants in the matrix more quickly and their distribution between the quadrants changed continuously. One of the most significant differences compared to the classic model was that cash cows in a phase of intense change have a lower profit share.

3) The need to constantly adapt the value proposition

Marketing management's ultimate objective is to create value for customers. Under the pressure of the incessant progress of digital technologies, what prospective customers consider as a value changes often and quickly, and competitors continuously think up new ways to satisfy changing expectations.

We have already seen how many times in the automotive industry a company has drastically lost the capacity to compete or even failed as a result of insisting on offering what had brought it success in the past and ignoring the evolution in demand. The accelerated change brought about by digital technologies obliges companies to constantly adapt their value proposition, that is, "how they serve their customer, what expectations they comply with, and at the end what value they deliver" (Rogers 2016).

The most effective response for marketing management is to place consumer expectations at the center of attention, that is, to adopt what is called a "consumer-centric attitude", quickly identify opportunities that emerge, and constantly adapt the value the firm delivers. By giving a new configuration to the value proposition, it is also possible to identify new prospective customers and new applications for existent products and services.

To better understand the strategic concept of value proposition, Rogers (2016) suggests comparing it with the four most common interpretations of market value. There are four strategic concepts: product; customer; "use case"; and "job to be done". The author then gives the example of the automotive industry, to demonstrate how these strategic concepts must be applied to decisions and planning, and how the value proposition is particularly useful "when you face the challenges of adapting and evolving your value to customers in response to changing needs and new opportunities posed by technologies".

4) New organisational structures and culture
To ensure that digital transformation has a profound effect and can remove resistances, the organisation of companies not "born digital" also needs to change. Many experts consider hierarchical structures and traditional values to be unsuited to fostering change towards a "digital culture".

In 'Cracking the Code of Change', an article that came out in the *Harvard Business Review*, Beer and Nohria (2000), after recalling that most programmes geared at changing the corporate culture fail, and that "every business change initiative is unique", identify two archetypes, or theories, of change. The two archetypes are based on different assumptions about why and how changes should be made. Theory E is change based on economic value. Theory O is change based on organisational capabilities.

The two authors caution that both theories are valid, reach some of management's goals, but have "often unexpected" costs. Theory E change strategies acquire the titles of specialised press, have a "hard" approach, use economic incentives to achieve objectives, drastic playoff, downsizing, and restructuring. Shareholder value is considered as the only legitimate measure of success. Theory O, on the other hand, has a "soft" approach to change. The organisation would be threatened if maximising shareholder value were the primary objective. Change is achieved, instead, by developing the corporate culture through "individual and organisational learning". To demonstrate how sharply Theories E and O differ, Beer and Nohria compare them on the basis of six dimensions of corporate change: goals, leadership, focus, process, reward system, and the use of consultants.

It is not possible to go into details here, but the analysis reveals that Theory O is the most suited to promoting change towards a "digital culture" of the organisation. It is based on the development of organisational capabilities, to encourage participation from the bottom up, experiment and evolve, and the focus is on the behaviour and attitudes of employees. Companies get ahead by giving employees more autonomy, fewer bosses, and the freedom to fail.

Among the companies, partnering with or buying startups seems to be considered as a way to accelerate the "digital culture". The process of change towards a new culture is not simple. Each manufacturer has its own roadmap. Bosch, which is rapidly adapting to the digital age, takes an "inside-out and outside-in" approach to fostering new ideas. To help make the transition and accelerate innovation, the automotive supplier is fostering a startup culture in its work force and a very customer-centric approach to innovation.[4]

To encourage the consolidation of a "digital culture", changes are also required in the composition of the supply chain. Daimler is among the automakers looking to establish a new structure of smaller partners and suppliers to tackle both the greater technological complexity of vehicles and the competition from companies such as Google, Uber, and Apple.[5]

4 The Long and Windy Road Ahead

"We always overestimate the change that will occur in the next two years and under-estimate the change that will occur in the next 10. Don't let yourself be lulled into inaction", said Bill Gates.

Now that cars are becoming smart devices, will established automakers be able to take a large share of the economic value of their industry? This is the question many are asking themselves.

Many drivers for change place pressure on the automotive industry. They are powerful and have still not demonstrated the full extent of their effects, but the automotive industry will undoubtedly undergo significant and continued change in the coming years. It is highly probable that within ten years, demand, the structure of competition and the players involved will be very different from the situation today.

History teaches us that companies in the automotive industry are slow to respond, but have repeatedly succeeded in facing threats from the environment (i.e., the sky-rocketing increase in petrol prices beginning in 1970 and the 2008 financial crisis). Some have been forced to give up (i.e., Saab, Fisker) or were absorbed by larger groups (Volvo, Rover Group and its Jaguar and Land Rover brands). Those who survived and prospered followed one key rule: flexibility, which means being ready to change their business model. Those who failed did not listen to the consumer or keep an eye on environmental trends. Established automakers have based the strength of their organisational structure on decades of experience in product development, on the stability of the "dominant design" of the product itself, and on established dealer networks. These strong points compensated for the traditional rigidity of large-scale

---

[4]"We have to increase speed in our innovation and become very customer-centric from the beginning" said Robert Bosch, CFO of Straub. "It's not just about buying startups, but having cooperation", said Straub. "We co-develop, we have R&D contracts, and often we are mentors." We look at everything from ideation to scaling of the business. Teams are encouraged to experiment with ideas and not be afraid to fail or pivot. We want them to learn how to fail early and do it with very little money. Bosch is also turning outward—partnering with startups, established companies and universities through hackathons and accelerators to develop new products and tap into emerging technologies. Source: "Bosch fosters a startup culture", *Automotive News*, 31 July 2018.

[5]"We have to talk about a new structure of partners and suppliers," said Sajjad Khan, vice president for digital vehicle and mobility at Daimler. "Nowadays, we have to handle agile startups, small software-driven companies or even individuals with innovative ideas." "Daimler looks to startups, small suppliers for innovative ideas". Source: *Automotive News*, 3 January 2017.

organisations in adapting to changes in the environment and fully understanding the extent of incumbent threats.

Eyes on a very difficult future. This time, the situation is different. The digital transformation is impacting an industry with product cycles lasting from three to five years, among the lowest average sector profitability levels, new lean, fierce competitors from other industries, and customers whose buyer behaviour is changing. To survive, automakers will need to master the digital transformation, but the uncertainty and the lack of a clear, structured vision of digitisation make it difficult to plan. In the absence of perfect foresight, major automotive manufacturers have turned to the principles or rules that have proven unbroken over time: embracing change and at the same time rendering existing approaches more efficient (i.e., VW has accelerated its margin market ambitions from 4% in 2018 to 6% by 2022) (Winkelhake 2017). There are four vital factors to observe for marketing management.

First. **Complexity and variety of demand.** Worldwide, with the rise in average incomes, the demand for mobility and thus for cars also rises. In the coming years, the 100 million cars produced in 2018 will undoubtedly be surpassed. Not everyone gets a share of the greatest wealth in the world, but far more people today have a higher standard of living than in any other period in history and, in spite of periodic recessions, wealth continues to grow faster than in any other previous time. Trends nonetheless differ depending on the geographical area and average income levels.

- In industrialised countries, the digital transformation is giving rise to a structural change in demand that is not due to temporary factors. Production is reaching the "peak car era", the stage in which growth should cease. The automotive industry is accelerating toward a time when private vehicle sales across the richest countries will plateau before making a descent. Bill Ford declared that, "[his] company is preparing for a future in which fewer cars are sold and people have access to a broader range of mobility solutions".
  The share of the world's population that lives in large cities is growing rapidly, generating major implications for private vehicle ownership and driving. In urban areas in industrialised countries, also as a result of traffic congestion and the limits that many governments have placed on vehicle use, the demand for mobility services is rising, especially among young people. Car ownership is being partly replaced by car usage through ride-hailing and car-sharing. Many people are convinced that the current urban model of privately-owned vehicles sitting idle up to a high proportion of the time is not sustainable as urban areas increase in size.
- In emerging countries, personal car ownership remains the main motivation to buy, as in industrialised countries outside the large cities; China and India are following different paths, but both are affected by strong traffic congestion and pollution in megacities (cities with over 10 million inhabitants).
- Faced with a complex, varied demand, companies with a large global presence must adopt a broad variety of business models and marketing management models. The idea of global marketing management (the same strategy for different geographic

markets) has long been surpassed, but the differences between the markets of different countries are now exasperated by the digital transformation (+).

Second. **Value is migrating from hardware to software.** In addition to their traditional competitors, established automakers must face the competition of non-traditional participants in the automotive industry. They include competitors with considerable resources (i.e., Apple, Amazon, and Google) and others that are lean and not weighed down by large assets (i.e., Uber, Lyft, Didi). They seek to disrupt the business model of the automotive industry through the introduction of new technologies, new products or services, and above all new business models.

Part of the value created by car production has already migrated from the mechanical side to software, and the trend is now accelerating; Morgan Stanley predicts that in autonomous vehicles, 40% of the value of an automobile will come from hardware, 40% from software, and 20% from the content that flows into the vehicle (advertising, news, games, park reservation, and others). In terms of software, Apple and Google have a great advantage over established carmakers. Waymo (the Alphabet Group, against Google) started a robotaxi service in Arizona in December 2018.

Third. **At a crossroads with business models.** Established players respond to the new demand and new competition by embracing the new. Most of them combine traditional "more vehicle-oriented" business with new "more mobility service-oriented" business. If car manufacturers are present in several geographical markets (industrialised and otherwise), they must maintain both business models, with the consequent organisational and decision-making difficulties.

In industrialised countries, the difficulty increases further when it comes to competing not only in the markets of traditional products (ICEs), but also in new markets: mobility services, connecting services, autonomous vehicles, and electromobility. This entails managing several business models, in most of which it will only be possible to make profit in the long term, while (in the markets of industrialised countries), traditional business sales (ICEs) will drop. As car drivers switch from ownership to services (ride-hailing and car-sharing mainly), revenues from sales will fall. It is unclear how long it will be before the demand for robotaxis and car-sharing will surpass that of car ownership. However, it is commonly believed that as soon as this occurs, the decline in ICEs will be fast. In the phase preceding the start of the breaking point, experts foresee a strong pressure on prices for ICE products because overcapacity increases and individual buyers are largely replaced with fleet buyers, who can obtain low prices.

Fourth. **Volatility is here to stay.** Under the pressure of the digital transformation, established carmakers must make broad adjustments to marketing policies and strategies. They must defend "old" business segments and open new business segments. It is interesting to observe how, in a phase of strong growth driven by technological innovation, an old tool like SWOT remains efficient. What opportunities are there? What threats? What strong points? What weak points? Understanding where we are going and remaining flexible is the matra, from brand positioning to segmentation, to the 7Ps.

In *Differentiate or die. Survival in our era of killer competition*, Jack Trout (2008) wrote that in brand positioning, being different means "differentiating yourself in the mind of your prospect" and amply illustrated the difficulties of positioning in "the battle for the customer's mind".

We can detect these difficulties nowadays in marketing management faced with the digital transformation. They are amplified and expanded by the fact that, "In the digital economy, customers are now facilitated and empowered to evaluate and even scrutinize any company's brand-positioning promise", as Kotler et al. observe (2017). Brands play a minor role, especially in mobility services.

Premium automakers, in particular the German brands Audi, BMW, and Mercedes Benz, have resources, history, and experience to affirm their brands. For volume manufacturers (low-end), this will be a hopeless task, however, since they are already nearing commoditisation today. In any case, even for premium bands, the future is hard. Their reputation was built on superior engineering, internal combustion engines, and driving enjoyment. These advantages are threatened by the change underway. Electrical motors are largely standardised and cannot command the same reward as ICEs. When vehicles drive themselves, they are unlikely to offer the same pleasure to drivers as in the past. BMW, which has advertised its cars as "The ultimate driving machine", must change its message.

German manufacturers reply that the engine is only part of the package, which includes smooth suspension, superior interiors, and high-level design. At BMW, they are convinced that it is possible to create a superior driving experience even with an electric engine, and that it is possible to offer a better driving experience with the management of electronics, even if the driver is no longer at the wheel. BMW has been manufacturing its own electric engines for some time now.

In electromobility, some established manufacturers have already clearly positioned their offer. Nissan Renault is in front of everyone with Leaf; Volvo has announced that, from 2020, all its new vehicles will be electric; and Porsche and Maserati have both announced EV vehicles. It is very likely that established carmakers will be able to establish their brands, also with EVs.

It should, however, be difficult for them to position themselves firmly in mobility services, given the commodity nature of transport over short distances in cities. Daimler with Car2Go, BMW with Drive Now, and VW with MOIA are already present with their own brands in these services. Even for start-ups (Uber, Lyft), it is a difficult task to build customer loyalty in mobility services. They have the advantage, however, of being lean and easily able to offer a wide range of mobility services without damaging their own image (a risk that car manufacturers take). Uber plans to become a one-stop shop for any kind of transportation, from cars and bikes, to trains and buses. Uber and Lyft already have their own scooter fleets in many cities. Their success highlights how electric scooters (and bike-sharing rentals) have started to gradually wear away car-hailing services in providing shorter trips in urban centres.

In connected services, compared to established manufacturers, third-party operators are in an advantageous position, given their greater flexibility in producing and updating contents. Lastly, those in control of software (Apple and Google) also dominate in the driverless industry. As we saw in previous chapters, they would have to give up on being vehicle manufacturers but would likely acquire the majority of the value created, providing mobility.

- Segmentation. This, too, is highly complex, and carmakers are faced with difficult choices. Alongside the traditional segmentation criteria, it is necessary to consider the preference for types of vehicle and autonomous driving, mobility services, and electromobility must also be included. For carmakers with a global scope, segmentation becomes even more complex. Indeed, each large geographical area follows its own pathway. In China, the restrictions on ICEs are very stringent, while in India they are restricted to very small areas in certain large cities. China is in a leading position in terms of EVs on the road, while in India they are almost non-existent.
A form of segmentation proposed by Winkelhake (2017) combines Nafta, the EU, and industrialised East Asia on the one hand, and the BRIC countries (Brazil, Russia, India, and China) on the other. The Nafta, EU, and industrialised East Asia markets are then divided into different mobility types with respect to mobility consumption: the Greenovators. These include High-Frequency Commuters, Silverdrivers, the Family Cruiser, and Low-End Users. Together, the first three segments make up around 75% of the market. The Greenovators are particularly interested in small vehicles, with the most advanced low-emission technologies, and electric vehicles. High-Frequency Commuters are daily commuters. When buying, they are interested in safety, efficiency, and reliability. The BRIC car markets are less heterogeneous and are primarily divided into the Basic segment (first-time buyers) and the Smart segment (second-time buyers).
- New role of the 7Ps. The difficulties of adapting the 7Ps have already been mentioned in Chap. 17. The digital transformation reinforces data analysis and can offer a better understanding of customers' needs and expectations. New products can be created or existing offers extended (an ecosystem with a focus on the consumer), which can foster personalised product offers. In terms of pricing, this offers greater transparency and advantages for customers. In relation to distribution, it results in a fall in sales volumes through dealers, while increasing internet sales for low-priced vehicles, but not for premium ones. Carmakers facing the challenge of digital transformation must adopt a multichannel distribution approach (through dealers, directly via the internet or through commercial platforms). How and what they sell changes. As well as vehicle sales, there are also structures for selling connected services and digital services.

To the question "How is selling mobility services different to selling vehicles?", the head of PSA Group's new mobility unit responded: "We are now involved in a totally new environment for us. Mobility services mean a 24/7 customer relationship, with round-the-clock availability and a seamless experience for highly demanding customers."[6]

References

Beer M, Nohria N (2000) Cracking the code of change. HBR's 10 must reads on change. 78(3):133–141

Chaffey D, Chadwick E (2016) Digital marketing. Strategy, implementation and practice. Pearson

Dodson I (2016) The art of digital marketing. Wiley

Foster R (1986) Innovation: the attacker's advantage. McKinsey & Co.

Foster R, Kaplan S (2011) Creative destruction: why companies that are built to last underperform the market - and how to successfully transform them. Crown Business

Jansson J, Andervin M (2018) Leading digital transformation. DigJourney Publishing

Kotler P, Kartayama H, Setiawan I (2017) Marketing 4.0. Moving from traditional to digital. Wiley

Leal-Millán A, Roldán JL, Leal-Rodríguez AL, Ortega-Gutiérrez J (2016) IT and relationship learning in networks as drivers of green innovation and customer capital: evidence from the automobile sector. Journal of Knowledge Management 20(3):444–464

McGrath RG (2013) The end of competitive advantage. Harvard Business School Press

Porter ME (2001) Strategy and the internet. Harvard Bus Rev 62–78

Porter ME, Heppelmann JE (2014) How smart, connected products are transforming competition. Harvard Bus Rev 92(11):64–88

Perkin N, Abraham P (2017) Building the agile business through digital transformation. Kogan Page Publishers

Rogers DL (2016) The digital transformation playbook. Columbia Business School

Ryan D (2015) Understanding digital marketing: marketing strategies for engaging the digital generation. Kogan Page

Simoudis E (2017) The big data opportunity in our driverless future. Corporate Innovators

Trout J (2008) Differentiate or die. Survival in our era of killer competition. Wiley

Venkatraman V (2017) The digital matrix. New rules for business transformation through technology. LifeThree Media

Winkelhake U (2017) The digital transformation of the automotive industry. Springer

---

[6]"PSA exec outlines risks and rewards from leap into mobility services". *Global Monthly*, September 2017.

CPSIA information can be obtained
at www.ICGtesting.com
Printed in the USA
LVHW080150100920
665503LV00006B/15